The Asian Century, Sustainable Growth and Climate Change

The Asian Century, Sustainable Growth and Climate Change

Responsible Futures Matter

Edited by

Moazzem Hossain

Griffith Business School and Griffith Asia Institute, Griffith University, Australia

Tapan Sarker

Griffith Business School and Asia Pacific Centre for Sustainable Enterprise, Griffith University, Australia

Malcolm McIntosh

Director, Asia Pacific Centre for Sustainable Enterprise, Griffith Business School, Griffith University, Australia

Edward Elgar

Cheltenham, UK • Northampton, MA, USA

Published by
Edward Elgar Publishing Limited
The Lypiatts
15 Lansdown Road
Cheltenham
Glos GL50 2JA
UK

Edward Elgar Publishing, Inc.
William Pratt House
9 Dewey Court
Northampton
Massachusetts 01060
USA

A catalogue record for this book
is available from the British Library

Library of Congress Control Number: 2013931631

This book is available electronically in the ElgarOnline.com
Economics Subject Collection, E-ISBN 978 1 78100 575 0

MIX
Paper from
responsible sources
FSC® C013056
FSC
www.fsc.org

ISBN 978 1 78100 574 3

Typeset by Servis Filmsetting Ltd, Stockport, Cheshire
Printed and bound in Great Britain by T.J. International Ltd, Padstow

Contents

Contributors

Ranajit Chakrabarty, Calcutta University, Kolkata, India

Smwarajit Lahiri Chakravarty, Calcutta University, Kolkata, India

Lai-Ha Chan, University of Technology, Sydney, Australia

Moazzem Hossain, Griffith University, Brisbane, Australia

Munshi Israil Hossain, University of Queensland, Brisbane, Australia

Shanawez Hossain, Asian Development Bank Institute, Tokyo, Japan

Paul Howard, Griffith University, Brisbane, Australia

Ronald C. Keith, Griffith University, Brisbane, Australia

M. Adil Khan, University of Queensland, Brisbane, Australia

Malcolm McIntosh, Griffith University, Brisbane, Australia

Vikram P. Murthy, Auckland University of Technology, New Zealand

Tapan Sarker, Griffith University, Brisbane, Australia

Vigya Sharma, University of Queensland, Brisbane, Australia

Patricia Short, University of Queensland, Brisbane, Australia

Clem Tisdell, University of Queensland, Brisbane, Australia

Yenny Tjoe, Griffith University, Brisbane, Australia

Preface

In January 2012, an international workshop was held at Griffith University's Asia–Pacific Centre for Sustainable Enterprise (APCSE) to brainstorm the ideas around the theme of the twenty-first century as 'the Asian century'. The aim of this workshop was to identify the major areas of policy shifts needed now to make the Asian century a reality. The workshop was attended by invited speakers with expertise on Asia's socio-economic and political developments, and subsequently they were approached to develop chapters for this volume. After going through a review process the respective papers were selected for publication, ranging from the examination of demographic conditions in Asia to mining resources and their potentials in the future. More than ten nations have been investigated.

This edited monograph was made possible by the financial assistance extended by APCSE of the Griffith Business School. We would like to acknowledge the support extended by the Director of APCSE, Professor Malcolm McIntosh. All of the chapters have undergone a rigorous review process by experts in the field. In this regard, the assistance by Professor Tom Nguyen, Adjunct Professor Adil Khan, Professor Peter Zingle, Professor Yukinobu Kitamura, Professor Jem Bendell, Adjunct Professor Harun Rashid and Dr Karen Vella is highly appreciated. Valuable logistical support was extended to us by Karen Brindley. We appreciate her hard work and commitment towards the holding of the International Workshop on 10 January 2012 which provided the basis of this volume.

Robyn White of the Department of International Business and Asian Studies of the Griffith Business School applied her usual publishing skills to fine-tune the draft and format the manuscript for Edward Elgar Publishing Limited. We are grateful to Robyn for her unconditional commitment to this project. Finally, we would like to congratulate the contributors who took time from their busy schedules to participate in the project, and thank them for their full cooperation in meeting various deadlines over the course of more than twelve months.

Moazzem Hossain
Tapan Sarker
Malcolm McIntosh
Brisbane, November 2012

Abbreviations

ADB	Asian Development Bank
APBD	regional budget plan
APCSE	Griffith University's Asia–Pacific Centre for Sustainable Enterprise
ASEAN	Association of Southeast Asian Nations
ASEAN+3	ASEAN and Brunei Darussalam, Japan, Republic of Korea
BDT	Bangladesh taka
BIMP–EAGA	Brunei Darussalam, Indonesia, Malaysia, Philippines – East ASEAN Growth Area
BIMSTEC	Bay of Bengal Initiative for Multi-Sectoral Technical and Economic Cooperation
BMET	Bureau of Manpower, Employment and Training
BOO	build, own, operate
BOOT	build, own, operate, transfer
BoP	base of pyramid
BOT	build, operate and transfer
BPL	below poverty line
BRIICS	Brazil, Russia, India, Indonesia, China, South Africa
CAREC	Central Asia Regional Economic Cooperation
CCP	Chinese Communist Party
CLTS	community-led total sanitation
CMDGs	Cambodian Millennium Development Goals
CSR	corporate social responsibility
DAK	Specific Purpose Grant
DAU	General Allocation Fund
DEPLU	Ministry of Foreign Affairs
DPR	People's Legislative Assembly of Republic of Indonesia
DPRD	*Dewan Perwakilan Rakyat Daerah* (Regional People's Legislative Assembly)
EAS	East Asia Summit
ERW	explosive remnants of war
FAO	Food and Agriculture Organization
FDI	foreign direct investment

FSM	Federated States of Micronesia
GDP	gross domestic product
GFC	global financial crisis
GHGs	greenhouse gases
GNI	gross national income
GONGOs	government-organized NGOs
GRDP	gross regional domestic product
GRI	Global Reporting Initiative
HBV	hepatitis B virus
HCR	head count ratio
HDI	Human Development Index
HH	households
HIV/AIDS	Human immunodeficiency virus/acquired immunodeficiency syndrome
HRS	household responsibility system
HYV	high yielding variety
ICMM	International Council of Mining and Metals
IDE	International Development Enterprises
IDU	intravenous drug user
IGOs	intergovernmental organizations
IITM	Indian Institute of Tropical Meteorology
IMF	International Monetary Fund
INPRES	Presidential Decree Fund
IPCC	Intergovernmental Panel on Climate Change
ISO	International Organization for Standardization
JICA	Japan International Cooperation Agency
LDCs	least developed countries
LICs	low income countries
MDGs	Millennium Development Goals
mmt	million metric tonnes
MPCE	monthly per capita expenditure
MPI	measuring multidimensional poverty index
MRP	mixed reference period
Mt	metric ton
Mtoe	million tons of oil equivalent
NBT	West Nusa Tenggara
NCIP	National Commission on Indigenous Peoples
NGO	non-government organization
NIEs	newly industrialized economies
NREGS	Mahatma Gandhi National Rural Employment Guarantee Scheme
NTT	East Nusa Tenggara Timur

NZ	New Zealand
OBC	other backward castes
OECD	Organisation for Economic Co-operation and Development
OL	outsourcing license
OLS	ordinary least squares
PDS	public distribution system
PEAC	Pesticide Eco-Alternatives Centre
PIF	Pacific Island Forum
POPs	persistent organic pollutants
PPP	purchasing power parity
PRC	People's Republic of China
PRSPs	poverty reduction strategy papers
QoL	quality of life
RACHA	Reproductive and Child Health Alliance
RCS	Resource Centered Strategies
RELA	*Ikatan Relawan Rakyat Malaysia* (Malaysian People's Volunteer Corps)
RPG	regional public goods
RRT	resource rent tax
SA	social accountability
SAARC	South Asian Association for Regional Cooperation
SARS	severe acute respiratory syndrome
SASEC	South Asia Subregional Economic Cooperation
SCEP	study of critical environmental problems
SCs	scheduled castes
SDO	Regional Autonomy Subsidy
SECSCA	Subregional Economic Cooperation in South and Central Asia
SIDS	small island developing states
SLO	social license to operate
SOEs	state owned enterprises
STs	scheduled tribes
TAR	Third Assessment Report
TKPK	*Tim Koordinasi Penanggulangan Kemiskinan*
TSRWSSP	Tonle Sap Rural Water Supply and Sanitation Project
TVEs	town and village enterprises
TWh	tera-watt-hour
UN	United Nations
UNCSD	United Nations Conference on Sustainable Development
UNDP	United Nations Development Programme

UNESCAP	United Nations Economic and Social Commission for Asia and the Pacific
UNFCCC	United Nations Framework Convention on Climate Change
UNFPA	United Nations Population Fund
UNICEF	The United Nations Children's Fund
USAID	United States Agency for International Development
VAT	value-added tax
WB	World Bank
WHO	World Health Organization

Introduction

Moazzem Hossain

The massive development in information and communications technology, together with economic reforms in the developing nations over the last two decades, have made responsible futures research more urgent and certainly more challenging. The new opportunities for responsible futures research in Asia are wide for scholars (Lele and Goswami 2011; Hean 2012; Desker 2012). It is clear that almost all developing nations of Asia have been achieving high to moderate growth over the last two decades, thanks to economic globalization and the information revolution, which have created more opportunities for trade in this region. Now is the right time to identify new areas of research with a view to address Asia's development challenges of the twenty-first century, which is also widely known as the Asian century.

Before illustrating the major purpose of the present volume, let us provide a brief background on the Asian century. The nineteenth century is regarded as the 'British' century, the twentieth was American, and now the twenty-first is called the Asian century. The debate on this subject goes back to 1985 when the US Senate Committee on Foreign Relations discussed it at length. More recently, in 2011 the Australian government established a committee on the subject of the Asian century, which examined the opportunities of the Australian economy in the future, with the Asian economies being predicted to reach the present day living standard of Europe by 2050 (White Paper 2012). However, the front runner in studying this phenomenon is none other than the Asian Development Bank (ADB) which published a report entitled 'Asia 2050 – realising the Asian century' in June 2011. The ADB was not hesitant in promoting this study worldwide, and arranged the launching of the report in Singapore, New Delhi and New York during the second half of 2011.

The ADB emphasized that Asia's rise in this century would be led by seven countries: India, China, Indonesia, Japan, Korea, Malaysia and Thailand. In 2010, these seven economies had a total population of 3.1 billion (78 per cent of Asia) and a gross domestic product (GDP) of \$14.2 trillion. But by 2050, the seven economies alone could account for

45 per cent of global GDP. Thus, under the Asian century scenario, Asia's GDP would rise to more than $170 trillion and would account for 51 per cent of global output compared to 27 per cent at present. In purchasing power parity (PPP) terms, GDP per capita in Asia would also rise to almost $46 000, compared to the global average of $36 600. Global economic activity would shift towards Asia and contribute to half of global trade and investment. About 3 billion more Asians would enjoy the same prosperous standard of living as present day Europeans (ADB 2011).

Indeed, if realized, there will be massive change by 2050 for Asia. This is not the first time that such an optimistic view has been seen on Asia. The World Bank in its East Asia Miracle study was very optimistic about the newly industrialized East Asian nations in the early 1990s. In 2002, an ADB study revisited the East Asia Miracle and came to the conclusion that, 'the future is not going to be a replay of the past' (Quibria 2002).

In 1968, before the 'miracle' took place on the Asian stage, the Swedish development economist Gunnar Myrdal published a three-volume best-seller called *Asian Drama: An Inquiry into the Poverty of Nations*. In the 1970s and 1980s, this was a must-read book for the students of development worldwide. Professor Gunnar Myrdal (a Nobel Laureate in economics in 1974) was also instrumental in providing policy alternatives to Scandinavian nations to adopt a social-welfare led growth model for sustainable development. As a result, these nations have been enjoying a long 'growth with equity' era over the last three decades.[1]

What did Myrdal have in mind for Asia? He was very pessimistic indeed, and emphasized that Asia was plagued by over-population, corrupt regimes and poverty, and risked being a region of failed nations (Myrdal 1968). The publication of these volumes was painful for Asian leaders. However, if one goes back to pre-1968 Asia, was Myrdal wrong? For all the present miracle and Asian century optimism, Asia indeed remains over populated and unable to free itself from corruption, and, of course, suffers from hunger and poverty in the early part of the twenty-first century. Unlike in Myrdal's time, Asia is now enjoying the information technology revolution, although it has been in two financial crises in the last 10 years and has the added burden of global warming induced climate change. Indeed, this is a time to revisit *Asian Drama* with a view to realizing or otherwise, the Asian century.

In order to challenge Myrdal in the present time, it is important that developing Asia keeps growing sustainably. In doing so, Asian nations need new policy initiatives under changed conditions to attain sustainable growth, with the additional major burden of global warming induced extreme weather conditions. As a result, policy interventions in new areas are needed with a view to bringing innovative and effective reforms in the present global environment, which is hostile financially and environmentally.

First, it is important to identify where responsible strategies are required, such as issues in transition, climate change, adaptation and mitigation, population and social safety net, natural resources, and so on. Second, it is important to investigate the future policy reforms for sustainable growth:

- productive sustainability in agriculture and food security for all;
- public revenue and infrastructure including public health and sanitation;
- population, migration and intra-country movement of people;
- decentralization and devolution of administration; and
- a sustainable momentum in growth.

Three billion Asians achieving present day European living conditions in 2050 is certainly a big ask. However, it is not impossible, if the European style institutions for society, economics, and politics (trust and tolerance) can be established in the next 40 years. There is an urgent need to fix a few problems, particularly in the areas of poverty, social protection, taxation (revenue generation) and infrastructure building (public spending), including public health, rural sanitation facilities and climate change.

Recently, the Food and Agriculture Organization (FAO) chief, Jose Graziano Da Silva, mentioned that 'the Asia–Pacific region remains home to two out of every three of the world's hungry. Sixty-two per cent of the undernourished population of the world lives in this region. That means around half a billion people hungry; that is half a billion too many' (Da Silva 2012, p. 8). He identified three major challenges for the region to 2050:

- to abolish hunger and to ensure food for all is a human right;
- to continue momentum for agricultural production given climate change and urbanization; and
- to do it in an environmentally, socially and economically sustainable manner (Da Silva 2012).

While all the challenges listed above are not new, to manage them in a sustainable manner becomes a new agenda. In recent years, the threat of global warming induced climate change has driven the world towards sustainable growth, developing and developed nations alike. As far as the challenges of adaptation to and mitigation of climate change are concerned, this region is certainly taking new measures and initiatives (Hossain and Selvanathan 2011).

In fact, Asia has been taking a leadership role in the world, in the area of adaptation to climate change. Huq (2012) claims that South Asia, due to its geography and the speed of urbanization, has a prospect of bearing the major brunt of climate change. He emphasizes that:

> The Asia–Pacific region has by far the biggest population as well as the largest number and variety of countries of all the world's continents. This variety includes developed countries like Japan, least developed countries (LDCs) like Bangladesh, small island developing states (SIDS) like the Maldives as well as the giants like China and India. At the same time, it has all the vulnerable ecosystems like mountains in the Himalayas, low lying coasts, small islands, mangroves and other forests, dry lands, major river systems like the Ganges, Brahamaputra and Mekong and perhaps most importantly, some of the fastest growing cities and mega-cities. (Huq 2012, p. 9)

With all these conditions, Asia indeed is placed in a unique position to face the challenges of global warming, more than other nations on the planet. The recent initiatives in preparing the region against climate change – such as governments' pro-active attitude towards adaptation, preparing the urban areas for adaptation, taking measures to create community-based adaptation approaches, making educational and research institutions hubs for gathering knowledge and management training – are reassuring (Huq 2012).

Thus, challenges for Asia to 2050 are manifold: social, economic and environmental. With all these challenges in mind, to keep growth momentum in Asia sustained for the next four decades is indeed a huge task. Moreover, some legitimate concerns can be found in the development literature: are high to moderate growth regimes, witnessed over the last 30 years, desirable without taking care of welfare and equity? Is the conventional growth model sufficient in bringing Asia's bottom one-third under the folds of prosperity? How are social evils such as corruption, declining law and order, and lack of job security being addressed meaningfully under the current growth model? In view of the above, the present study attempts to identify and develop the major areas of future strategy and the ways to act responsibly to realize the Asian century. The corruption and law and order issues in Asia have been left outside the parameters of this volume since a comprehensive investigation of these areas needs an entire volume to itself.

The book has been presented in three parts, comprising three chapters in Part I, six in Part II and three more in Part III. A brief summary of each of the chapters is presented below for a ready reference to the reader.

The chapters under Part I identify the issues which remain as major threats to realizing the Asian century. Tisdell, in Chapter 1, highlights that the continuing contribution of China and Vietnam to the realization of the Asian century is going to be influenced (along with other factors) by how well their governments are able to deal with social conflicts involving the rights to property and its transfer. While early agricultural reforms were socially responsible, they stopped short of giving agricultural households full property rights in the land assigned to them. In Chapter 2, Hossain demonstrates that the continuing pressure from population remains one of the major hurdles in alleviating poverty in Asia. The populous nations of Asia, particularly South Asia (India, Bangladesh and Pakistan), the home of almost half a billion poor, have been falling further and further behind in achieving the poverty reduction targets of the Millennium Development Goals (MDGs). Hossain and Sarker claim in Chapter 3 that the prospect of the Asian century depends heavily on the ability of Asian countries to navigate effectively the resource constraints from social, political and economic points of view. Developing and emerging Asian nations appear to be more vulnerable to these constraints due to higher risks associated with climate change and lower adaptation capacity.

Part II captures the policy issues which matter most in realizing the Asian century. In Chapter 4, Sarker illustrates the issues of responsible taxation policy, which is regarded as one of the preconditions for Asia's inclusive prosperity. He provides an analysis of the term 'justice' in taxation and identifies a new paradigm for designing tax policy that incorporates considerations of fiscal sustainability, capital maintenance and sustainable economic development. Chakrabarty and Chakravarty, in Chapter 5, provide a quantitative analysis of India's agricultural growth, which is likely to be threatened by global warming in the future. They prescribe some policy changes in the areas of food and agriculture and in the rural non-farm sector in order to reduce rural poverty through transformation of rural areas. In chapter 6, Chan and Keith address the policy issues of China's public health. They observe that, while China has become more integrated with the world economy and international organization, this nation persists in adopting a state-led multilateral approach to all types of NGOs. New social organization has proliferated outside the formally exclusive monopolies of the Communist Party's mass associations. Health NGOs have been allowed to come through China's 'open door' policy but their entry is conditional on respect for Chinese sovereignty and Party control on public health delivery. In Chapter 7, Howard examines the predicament with Asia's sanitation facilities in general, and those of Cambodia in particular. Despite Cambodia's relatively low levels of access to sanitation, the potential for harnessing the efforts and energy of the various actors at all levels is a cause for optimism. As the developing economies in Asia continue to grow, coordination of policy at all levels of government will be necessary to develop sanitation infrastructure and facilities. A sustainable sanitation infrastructure which caters to all citizens' needs and health outcomes is an essential part of the vision of a truly inclusive 'Asian century'. Tjoe, in Chapter 8, shows that decentralization reform in Indonesia has enhanced poverty reduction, and also demonstrates that the decentralization policy was adopted effectively as a pro-poor policy in this nation. This pro-poor decentralization is important to ensure sustained poverty reduction in the future. What the new system needs to enforce is transparency, participation and policy coherence at the local level in order to eliminate inefficiencies. Chapter 9 by Hossain, Khan and Short demonstrates that Bangladesh's unskilled labour migration to Malaysia has brought for the country a major boost in remittances, and certainly adds hugely to the scarce sources of foreign exchange. It is not only skilled migration from Asia to the rest of the world that has been bringing prosperity to the region – intra-Asia unskilled migration is also playing a major role in the collaboration and economic integration of the nations of Asia, but raises issues of serious concern.

Part III identifies the major challenges facing Asia if it is to realize the prosperity under the folds of the Asian century. In Chapter 10, Sharma and Sarker demonstrate Asia's potential for a resource boom to help support prosperity and make the Asian century real. They examine Mongolia's recent rise in mining resources and its promise for the future. The challenges of a mining boom in the early twenty-first century in Mongolia are also elaborated. Murthy, in Chapter 11, demonstrates Asia's business challenges from the corporate responsibility, strategy and leadership point of view. He demonstrates that corporate sustainability is a significant strategic capability for business in developed and developing economies alike. Highlighting the initial findings from an ongoing neo-Glaserian longitudinal study of senior business leaders in Australia and New Zealand, Murthy proposes a Zeitgeist organizational leadership repertoire of virtues, practices and enactments for present and future high-velocity and complex environments. Finally, McIntosh presents the concluding chapter with an alternative thought on the Asian century phenomenon. According to McIntosh in Chapter 12 the idea of the Asian century depends where one starts and how one wants to see the world. But from any perspective it is myopic and self-referential and depends on what one wants to see – and perhaps where one wants to belong.

NOTE

1. There is a (possibly invented) story about a conversation between Gunnar Myrdal and Gustav Cassel, in which Cassel is reported as saying, 'Gunnar, you should be more respectful to your elders, because it is we who will determine your promotion', and he replied, 'Yes, but it is we who will write your obituaries'. It must also be noted that Gustav Cassel endorsed the 'purchasing power parity' (PPP) measure in the early 1920s. Myrdal was also unhappy about his colleague Professor Friedrich von Hayek who was awarded the Nobel Prize with Myrdal in 1974, due to Hayek's contribution to free market economics.

REFERENCES

ADB (2011), *Asia 2050 – Realising the Asian Century*, Manila: Asian Development Bank.

Da Silva, J.G. (2012), 'Towards a hunger-free Asia and Pacific', *The Daily Star*, Dhaka, 15 March.

Desker, B. (2012), 'Don't let the past shape Asia's future', Singapore Global Dialogue, S. Rajaratnam School of International Studies, National University of Singapore, Singapore.

Hean, T.C. (2012), 'Asia's rising, but faces severe challenges', Singapore Global Dialogue, S. Rajaratnam School of International Studies, National University of Singapore, Singapore.

Hossain, M. and E. Selvanathan (eds) (2011), *Climate Change and Growth in Asia*, Cheltenham, UK and Northampton, MA, USA: Edward Elgar.

Huq, S. (2012), 'Asia leads the world in adaptation to climate change', *The Daily Star*, Dhaka, 16 March.

Lele, A. and N. Goswami (2011), *Imagining Asia in 2030: Trends, Scenarios and Alternatives*, New Delhi: Academic Foundation.

Myrdal, G. (1968), *Asian Drama: An Inquiry into the Poverty of Nations*, New York: The Twentieth Century Fund.

Quibria, G.M. (2002), *East Asian Miracle Revisited*, Manila: Asian Development Bank.

White Paper (2012), 'Australia in the Asian century', White Paper, Prime Minister's Office, Australian Parliament, Canberra, October.

PART I

Future Issues of Asia

1. Agriculture, structural change and socially responsible development in China and Vietnam

Clem Tisdell

INTRODUCTION

It is well known that both China and Vietnam have achieved high rates of economic growth since they began their economic reforms. These reforms were designed to transform their economies from a high degree of dependence on central planning and command to ones much more reliant on market systems. Also these reforms were accompanied by policies to make their economies more open to the outside world and encourage foreign direct investment. Although China's economic reforms began in 1978, Vietnam did not commence its reforms until 1986 when it began its *Doi Moi* policy designed to renovate its economy. As a result of their economic reforms, the economies of both China and Vietnam have undergone tremendous structural change in a relatively short time-span (see, for example, Tisdell 2009b; Tisdell 2009a). As is to be expected, these changes have given rise to several social tensions. For example, land disputes have resulted in social unrest in some areas of China and Vietnam. In part, these disputes highlight problems in reallocating the use of land to accommodate structural economic change.

Possibly, the major structural change in these economies has been the decline of their agricultural sector relative to the remainder of their economies, especially secondary industry. These changes have been accompanied by a rapid increase in urbanization (and accompanying rural-to-urban migration) and a decline in employment in agriculture. These are trends normally associated with economic growth and development (Clark 1957). Probably, the economic reforms in China and Vietnam have had the greatest social and economic impact on agriculture and rural communities. Therefore, this chapter focuses on the socio-economic problems faced by agriculturalists in adjusting to structural change in these economies and considers to what extent economic adjustment policies adopted by the reformers have exhibited social responsibility in relation to agriculturalists, who in this case consist mostly of very small-scale farmers. These small-scale (mostly poor) farmers rely on family labour for their production and in many cases, supply a substantial amount of their own food from their own land.

The discussion in this chapter proceeds in the following way. The process of economic reform in relation to the agricultural sector is considered, including the development of non-agricultural industries in rural areas, town and village enterprises (TVEs) in China's case. Then the question of agricultural land rights and rural land disputes is raised. This is followed by consideration of the social and economic desirability of further structural change in agriculture, namely a change to greater commercial industrial-type agriculture and a reduction in small-scale family-based agriculture, which is seen by some Chinese and Vietnamese policymakers as of a peasant-like nature, backward and inefficient. The political push by some groups for the extension of commercial industrial-type agriculture in China and Vietnam has created social tensions in rural communities because of some land transfers for this extension. However, forced land transfers required to accommodate other structural social and economic change seem to have created even more social conflict.

Whether or not the policies that have been developed by China and Vietnam to deal with these problems are socially responsible is a matter for debate and will be considered here. As is analysed in the final part of this chapter, their policy of making gradual economic reforms, while socially responsible initially, has now resulted in social conflict and a political dilemma.

AGRICULTURE AND THE REFORM PROCESS

The main focus in this chapter is on China but mention will also be made of similar issues that have arisen in Vietnam as a result of structural economic change.

Economic Reforms Began with Agriculture

Both China and Vietnam began their market economic reforms with agriculture. Was this because it was the socially responsible thing to do or was this for other reasons? The answer is complicated.

China's economic reforms began with the introduction of the household responsibility system (HRS) in agriculture (Chai 2011, ch. 15; Tisdell 1993, ch. 8). Among other things, this involved the transfer of collectively owned and managed agricultural land, agricultural animals and equipment to farm families on a contract basis. Land was assigned to farm families for a fixed period of 30 years, but titles to the land were retained by the collectives (village governments) and were not given to individuals.

Chai (2011, p. 168) states: 'In order to minimise social, political or economic disturbances, China developed a unique approach to privatisation for it did not want to engage in the outright sale of state and collective properties. In the agricultural sector, the way to privatisation was filtered through a contracting-out system'. While this process or approach minimized social disruption initially and therefore, at the time, was socially responsible, subsequently (as will be discussed later) this method of allocating land became a source for rural land disputes and social unrest. This is because some local governments facilitated compulsory acquisition by developers of the land initially assigned to farm families. Similar developments can be observed in Vietnam.

It can be argued that it was socially responsible for China and Vietnam to begin their economic reforms with agriculture. At the time these reforms began, most Chinese and Vietnamese were engaged in agriculture and were much poorer than the urban population. They were therefore in greatest need of a boost in their incomes. There is no doubt that as a result of the introduction of the HRS, managerial efficiency in agriculture greatly increased compared with production under the commune/collective system and that allocative efficiency also improved. Liu Guang and Liang et al. (1987) stated:

> The contract responsibility system based on the household with remuneration linked to output has been introduced, which clearly defines the obligations, directly benefits the peasants, involves simple procedures and makes maximum use of the enthusiasm of the masses of the peasants by assuring them of their power to make decisions on matters relating to the management of production . . .

Consequently, these agricultural reforms helped to raise the incomes of Chinese farmers even though on average their incomes remained lower than those of urban dwellers.

Nevertheless, it seems likely that the exercise of social responsibility was not the only possible reason why China's economic reforms began with agriculture. Another important influence was almost certainly political realities or practicalities. Deng Xiaoping was restrained in his reformist ambitions by party elders who, according to Chai (2011, pp. 163–4), 'much preferred central planning and moderate reforms' to the modernization programme of Deng Xiaoping. To have begun the economic reforms in urban areas would have rapidly generated urban social strife given the vested interests of urban-dwellers in the existing system which favoured and protected them economically to a large extent. Also demonstrations and protests are more likely to occur in urban rather than rural areas when there is social unrest. The prudent political course, therefore, was to begin the economic reforms in rural areas where economic benefits were likely to be achieved quickly. At the same time, village governments or collectives would still be able to retain significant economic and political power, thereby ensuring their continuing support of China's Communist Party and the central government. For instance, local governments retained considerable political power over members of the collectives or village and their use of land. In turn, this contributed to the continuation of political cohesion.

Town and Village Enterprises

The Chinese government's support for the growth of rural industries (TVEs) early in its period of its economic reform was another important development which significantly benefitted farmers. In addition, it was a stepping stone to further extension of market reforms in China into urban areas (Tisdell 2009b). Once again, it can be argued that while this was a socially responsible strategy, it was also heavily influenced by practical political considerations. To have begun such reforms in major urban areas in which state owned enterprises (SOEs) dominated production would have courted immediate strong social opposition because the job security of most urban dwellers depended on such enterprises.

The encouragement of the development of rural non-agricultural enterprises by the Chinese government was socially responsible because this development provided non-agricultural employment possibilities close to home for farmers. This reduced the cost to farmers of switching from agricultural work to non-farm work compared to the alternative of seeking employment in a large city far away from home, a possibility anyway restricted by the operation of the Household Registration System. The importance and nature of the costs involved in Chinese farmers changing location to switch their employment is discussed and analysed by Tisdell (2009a). The development of non-agricultural rural enterprises (TVEs) facilitated a reduction in the amount of China's surplus agricultural labour by providing an alternative to employment in agriculture. After the introduction of the HRS, the surplus of agricultural labour may have increased or may have become more pressing, and therefore the need for political action to absorb this surplus may have become more apparent. Chai (2011, p. 169) states: 'Initially the development of TVEs was promoted by the government to soak up the surplus labour generated by the higher productivity which had resulted from the introduction of the HRS in the countryside. Later on they were promoted because they provided a competition to SOEs'.

The development of TVEs added to rural incomes because on average the productivity increased of those who switched from employment in agriculture to work in TVEs. Initially also TVEs were owned and controlled by rural collectives thereby strengthening their political power base. However, according to Chai (2011, p. 169), many TVEs have now been converted to private ownership. Chai does not discuss the mechanics used for the conversion of TVEs to private ownership but in several cases, assets may have been privatized at bargain rates.

Initially also the Chinese government favoured the growth of TVEs because it was worried about the social consequences of mass rural migration to large urban areas. Urban dwellers were hostile to such migration because they were fearful that it would create a social and economic burden for them.

Vietnam did not have China's system of TVEs. This may have hindered its structural economic adjustment. Now, however, Vietnam's government is giving greater attention to the development of industries in rural areas.

While the virtues of using TVEs as a part of China's process of gradual economic reform have been stressed by many writers, some limitations should be mentioned. To what extent, in the long run, are or were the activities engaged in by the TVEs economically viable and did the establishment of TVEs result in the best locations for industrial development in China? It seems likely that location of the non-agricultural economic activities engaged in by TVEs was not the most economically efficient nationally. Furthermore, in the longer term, many established TVEs might be unable to withstand increased market competition and therefore, eventually will no longer be economically viable. The purist economic ideal of optimal economic allocation of resources was probably not achieved by the development of China's TVEs.

However, this purist ideal is to a large extent immaterial because from a dynamic point of view the development of TVEs was a success. It provided rapid economic benefits to rural dwellers at no expense to urban dwellers. The policy was a stepping stone for the extension of China's market reforms to urban areas. According to Tisdell (2013), it is more important to consider economic processes rather than comparative static analysis in formulating economic policies.

On the other hand, as market competition becomes more widespread in China many of the TVEs established in earlier times are likely to be no longer economically viable. Therefore, some socially distressing future economic adjustment is likely to occur. However, China is in a better position than in the past to deal with this readjustment given the considerable economic growth experienced by it.

AGRICULTURAL LAND RIGHTS

While the gradual nature of China's and Vietnam's economic reforms can be interpreted as being socially responsible (although this gradualism was to a large extent dictated by political feasibility) because it minimized (early) social conflict, this gradualism also sowed the seeds for some future social conflict. It has resulted in claims that China's and Vietnam's socio-economic systems are not operating in a socially responsible way. As already indicated, as market competition in China increases some TVEs are coming under financial stress and they complain of unfair competition from SOEs and other more favourably placed business enterprises. This is reflected in a news item by Dong Wei (2012) which appeared in the *China Youth Daily* and which was edited and translated by the *People's Daily* Online and reproduced in the Travel Impact Newswire. This item complains about competition facing farmers and TVEs from SOEs.

A similar type of situation has developed since the introduction of the HRS in agriculture. Land belonging to the village communes or collectives was divided between families on an egalitarian basis for them to use it for agricultural production. Households were given the conditional right to use this land for 30 years. They could not mortgage or sell it, but in some circumstances, it could be rented. Titles to the land were retained by the village government and they were authorized to reallocate the land at any time in response to demographic and similar changes.

This may have seemed socially responsible at the time. This form of allocation of the land provided farm families with some social security and yet allowed scope for land reallocation as needs changed. Furthermore, by severely restricting the transfer of agricultural land, it preserved the rural social security base of farm families. Family members of farmers migrating for temporary employment in cities, China's floating population estimated by the OECD (2010) to consist of 74 million people in 2005, had a social blanket to which they could return should they become unemployed or otherwise disadvantaged after migrating.

In recent years, government restrictions on the transfer of agricultural land in China and Vietnam and on the use of agricultural land have been relaxed. Local governments retain property rights in agricultural land and have taken advantage of this relaxation and some ambiguities in the law. Chai (2011, p. 177) observes for China the following:

> Theoretically, village officials and township cadres cannot convert or lease farmland for other purposes without the majority consent of their villagers. In practice, however, to boost the local government revenues, land was [is] often leased to factories or converted to other purposes without due process and then sold to property developers at inflated prices.

A report in the *China Daily* updated on 6 November 2010 (Anon 2010a, p. 1) states that 'about 65 per cent of mass incidents in rural areas are triggered by land disputes which are affecting rural stability and development more than any other issue'. Agricultural land is being 'seized' to facilitate economic development needed to support China's economic restructuring. According to Professor Yu Jianrong of the Chinese Academy of Social Sciences, land disputes occur because of forced land acquisition and the low level of compensation paid to those who are forced to relinquish their land. Compared to the market price of the seized land and payments made by those acquiring it, a low amount is paid by village governments to households whose land is repossessed. Professor Yu Jianrong is quoted as saying 'Since the reform and opening up, more than 50 million farmers have lost their land and nearly all have no jobs or social insurance. This has caused social conflict' (Anon 2010a, p. 1).

The *China Daily* (Anon 2010b) also reported on 4 November 2010 that some local government officials have broken the law by taking over farmland from farmers and selling it. Breaches of the law are occurring on a large scale and this is a matter of concern for China's paramount leaders. Even when farmers have legal rights to object to such seizures, it is difficult for them to protect their rights, because they are poor. According to Wang Chunguang, a social mobility researcher at the Chinese Academy of Social Sciences, 'many farmers could not protect their rights because they were too poor to afford the cost of a lawsuit and that this explained why so many people appealed to higher authorities for help' (Anon 2010b). It also helps to explain why land disputes have resulted in so many mass incidents and demonstrations in rural China. How these problems can be solved in a socially responsible and equitable manner is unclear.

This is because legal solutions often fail to work, as most affected farmers are too poor to defend their legal rights. Furthermore, legal outcomes are uncertain and significant time may elapse before a decision is reached. Low levels of education of farmers could also contribute to the difficulties which they face in defending their rights. While arbitration may help to reduce the problems involved, even arbitration is not costless. Furthermore, farmers do not legally possess the land assigned to them as private property. Therefore, legally they are not entitled to full compensation at market value for the repossession of their land. The amount of compensation farmers should receive for land repossession has not been accurately determined but there is no legal requirement that it should be the full market value of the repossessed land. Whether or not farmers should have full private property rights in the land assigned to them is a different matter. However, they do not have these rights at present.

The agricultural land situation in Vietnam is similar to that in China. The repossession of agricultural land results in heated disputes and strong opposition from affected farmers. Bland (2012, p. 1) states in relation to the Vietnamese case of farmer Vuon whose land was forcibly taken:

> Vuon's case quickly became a *cause celebre*, as many sympathised with his frustration with greedy local officials and the lack of avenues for proper legal redress. With courts controlled by the ruling Communist party and land law unclear, there are plenty of opportunities for abuse by officials in league with developers.

In Vietnam, as in China, farmers hold conditional land use rights, and some trade in these rights is allowed but land titles are held by local governments. Furthermore, as in China, ambiguities exist in the laws relating to land rights and the power of individual farmers to protect their rights is weak in comparison to the power of local governments and the state take away these rights. According to Bland (2012, p. 2), the Prime Minister of Vietnam 'admitted that while the land laws have been improved since Vietnam began opening up its economy in the late 1980s, many of the hundreds of legal documents relating to land "remain unclear, even overlapped or conflicting"'.

The taking by the state (public bodies) of land held by private individuals or entities is a contentious issue in all countries. All governments have the right to take private property for public purposes and rules vary about the amount of compensation to be paid to the owners of that property (Tisdell 2009c, p. 126–7). However, the magnitude of the problem is much greater in transitional economies such as China and Vietnam where freehold or similar land titles are absent and land has continued to be owned by the state. While land reforms made as part of the market reform seemed to be socially responsible at the time, with the economic growth and structural change of China and Vietnam, they have become a major source of social conflict. The governments of both China and Vietnam are searching for ways to overcome these conflicts. However, they face a dilemma. The granting of full private property rights in agricultural land to households will weaken the power base of local officials (who are usually supporters of the ruling party) and indirectly this could weaken the political power of the central government. However, privatization of land might appease farmers and facilitate the transfer of agricultural land to alternative higher value users. On the other hand, these households selling their land may have less social security than before. Nevertheless, in China's case, its recent introduction of collective medical and pension insurance for rural areas will help to alleviate this problem (Chai 2011, p. 177).

SMALL-SCALE HOUSEHOLD FARMS VERSUS LARGER COMMERCIAL INDUSTRIALIZED FARMS

In both China and Vietnam, households (rather than business corporations) have the rights to use most of the available agricultural land. The area of land available to individual households is extremely small by Western standards. The 2000 Agricultural Census in Vietnam recorded that just under 69 per cent of agricultural households had less than 0.5 ha of land and that about 94 per cent had less than 2 ha of land (General Statistical Office of Vietnam 2007). In China, the average area of land held by agricultural households in 2008 was just a little less than 0.5 ha (Wu 2009). Farm households rely on their own families to provide labour (rather than on hired labour) and only partially rely on market exchange to meet their economic needs.

Some policymakers in both China and Vietnam have questioned whether reliance on peasant-like agriculture of this type is the most economic way to supply agricultural production. They advocate larger agricultural units which are organized completely on a commercial basis and which use 'modern' production techniques (for more information, see Tisdell 2010). It is believed that significant economies of scale can be achieved for some forms of agriculture, such as pig and poultry production. This may also be true for the cultivation of some crops such as cotton. China has large mechanized cotton farms in Xinjiang, although at the same time much of China's cotton is produced by households engaged in mixed agricultural production on small land holdings (Zhao and Tisdell 2011).

In both China and Vietnam, some policymakers favour giving economic concessions to companies that are willing to develop commercial large-scale agricultural production, particularly in the livestock sector, especially pig production. With rising incomes, the demand for livestock products has risen sharply in China and Vietnam and pork is an important component of the diet in both countries. This raises the question of whether it is socially responsible to try to accelerate the replacement of household agricultural production by commercial agricultural production. In the absence of suitable alternative employment for agricultural households, accelerating the process of agricultural commercialization

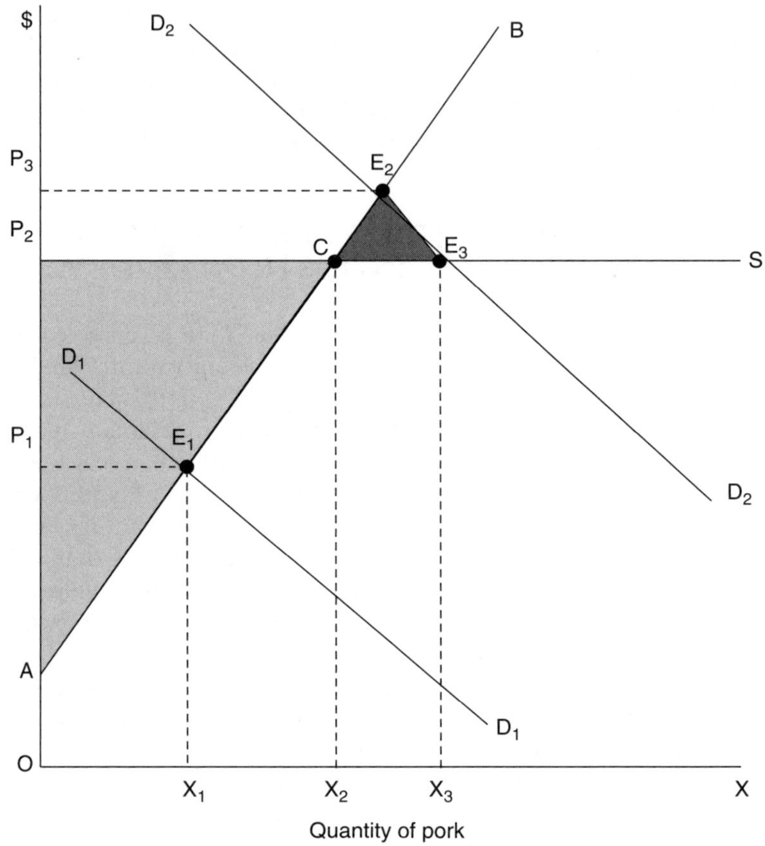

Figure 1.1 *Policies favouring the expansion of commercial agricultural production at the expense of household agricultural production can be socially irresponsible in transitional economies*

is likely to cause social distress for agriculturalists depending on small farm lots. At the same time, facilitating the trend towards commercialized agriculture without subsidizing it seems to be socially responsible.

Figure 1.1 can be used to discuss the situation. For illustrative purposes, let us suppose that pork production is being considered. Currently most pork in China and Vietnam is produced by agricultural households. Assume that the supply curve of pork by households in China or Vietnam is as indicated by line AB. If only households supply pork, the equilibrium price of pork is P_1 if the demand relationship is as indicated by the line D_1D_1, and if the demand curve for pork rises to D_2D_2, the equilibrium price of pork increases to P_3. Assume now that commercial producers could supply pork at a cost per cent equivalent to P_2. If they are free to do so and if households only lose a negligible amount of their land to make this possible, the supply curve of pork is now ACS. Commercial production of pork is not economic when the demand for pork is low (D_1D_1) but it is when the demand for pork is high (D_2D_2). Using the Kaldor–Hicks criterion, economic welfare is increased by permitting commercial production of pork when demand is high. While the economic surplus of agricultural households is reduced as a result of commercial production when the demand for pork is high by an amount equivalent to the area of trapezium $P_2P_3E_2C$, consumers' surplus rises by an amount equivalent to the area of trapezium $P_2E_3E_2P_3$ if commercial production occurs. Hence, the amount of economic surplus gained by consumers exceeds that lost by household producers by an amount equal to the area of the hatched triangle CE_3E_2.

Note that even when demand for pork is high, it is uneconomic to replace all household production of pork by commercial production. In the case illustrated, if this were done it would result in a reduction in economic welfare equivalent to the area of the dotted triangle ACP_2. Note that the supply curve of output by agricultural households should be based on their opportunity costs. One would expect that as the economies of China and Vietnam continue to grow, these costs would rise because of expanded opportunities for farm families to gain non-agricultural employment yielding increased income. Therefore, it will become more economic for a larger amount of agricultural production to be supplied commercially rather than by households. Nevertheless, in economic transition, both sources of supply are likely to be economic and it would be socially irresponsible to provide economic incentives which favour the expansion of commercial household production at the expense of agricultural household production.

DISCUSSION AND CONCLUSION

In most economics textbooks, economic policies are analysed by means of comparative statics and are divorced from their historical, political and institutional contexts. As the discussion in this chapter reveals, such an approach is inadequate. The implementation of economic policies involves dynamic processes and is shaped by the historical, institutional and political contexts in which they occur. Therefore, a dynamic approach based on political economy rather than pure economic theory (especially comparative statics) is preferable for the analysis of economic policies. It is only by adopting such an approach that one can adequately appraise China's and Vietnam's economic reforms in recent decades.

It is sometimes said that politics is the art of the possible. A gradual and measured pace of economic reform was possible in China and Vietnam beginning in rural areas, whereas initial rapid reforms based on altering the economic system within large urban settlements would have been a recipe for civil strife. The early reforms also seem to have been socially responsible but they set in motion processes that have subsequently caused social conflict and have led to accusations that government officials are acting in a socially irresponsible manner. This is quite evident in cases involving the taking away of the rights of farmers to use agricultural land, and it also arises in relation to the taking away of the use of land by urban households.

Because of economic growth and structural economic change set in train by China's and Vietnam's economic reforms, it has become necessary to alter the allocation of land to accommodate structural economic change. Whether or not this reallocation could best be achieved by establishing a system of complete private property rights in land and in capital is a moot point. Such a change would most likely weaken the power base of the ruling parties in China and Vietnam. Whether it would completely resolve individual grievances about land rights is unclear, as also is the extent to which it would facilitate land reallocation to accommodate structural change. Bargaining by individual landowners could hold up socially desirable changes in land use.

Ambiguities in the law and principal-agent problems (for a discussion of this problem see Anon 2012) play an important role in the occurrence of land disputes in China and Vietnam. On the one hand, village governments are supposed to be the agents of the local people but are also responsible to central authorities which represent, supposedly, the whole population of the country. However, both these principals (in theory the masters) have only limited means to control their agents (village governments) and the agendas of the agents can differ from those of the principal. This provides scope for local government officials to engage in actions which are seen by farmers as being socially irresponsible.

It was also observed that the development of TVEs in China seems to have been a socially responsible policy initially. However, the adoption of this policy is largely explained by political strategy. As the Chinese economy continues to grow and economic competition increases in China, it seems likely that many TVEs will no longer prove to be economically viable. Thus some social conflict is likely to be generated and further industrial restructuring is likely to be needed.

A further source of potential stress for rural households in China and

Vietnam is the development of large-scale commercial agriculture. Some policymakers favour policies which give economic concessions to companies willing to develop large-scale commercial agriculture, particularly in the livestock sector. The livestock sector is of central interest because the demand for livestock products in China and Vietnam has grown rapidly as a result of rising incomes and has forced up the prices of livestock.

It is pointed out that giving economic concessions to encourage the growth of large-scale commercial agricultural production does not seem to be socially responsible. The case for artificially accelerating the change from agriculture mostly based on households with small landholdings is weak from an economic point of view and does not pay enough attention to the low level of opportunity cost currently incurred by household members by continuing to engage in agriculture. As their opportunity costs rise (for instance because they can obtain a higher income by being employed elsewhere), economic processes will naturally ensure that a larger proportion of agricultural production is supplied by commercial enterprises. Forcing such structural change by government intervention is likely to be socially and economically damaging.

The continuing contribution of China and Vietnam to the realization of the Asian century is going to be influenced (along with other factors) by how well their governments are able to deal with social conflicts involving the rights to property and its transfer. While early agricultural reforms were socially responsible, they stopped short of giving agricultural households full property rights in the land assigned to them. At the time, this was most likely the only practical course of action for the Chinese and Vietnamese governments. However, with continuing structural change in their economies and the need to alter the allocation of the use of land, significant social conflict has emerged as a result of agricultural land being transferred to alternative uses by decision of village governments and officials. Another matter of continuing concern in China is whether all TVEs will remain competitive as China's market economy develops. In rural areas where TVEs are no longer competitive, dislocation of employment will occur. In both China and Vietnam, the extent to which the development of commercial large-scale agriculture should be encouraged is also a matter for continuing debate. Providing special government benefits for the development of this type of agriculture will damage the economic interests of agricultural households and could add to social unrest in rural areas. For the reason previously outlined, such favouritism would not be a socially responsible policy.

The above issues have arisen because significant economic growth and structural changes have been set in train by the gradual approach of China and Vietnam to their economic reforms. While such gradualism is socially responsible and causes little or no social conflict in the short term, it can sow the seeds for significant social conflict in the longer term as the experiences of China and Vietnam illustrate. At the same time, it is clear that all dynamic economies have social conflicts of varying magnitudes, the nature of which alters with the passage of time. For example, in India there are also concerns about the increasing commercialization of agriculture and the changing social situation of agricultural households. In China and Vietnam, however, the issues involve the very core of communism, namely the desirability of collective ownership of land and other means of production, and are a potential problem for the retention of political power by the ruling communist parties. Consequently, adjustment of property rights is difficult from an ideological and political perspective even though social stability is being put under stress by the present laws.

REFERENCES

Anon (2012), 'Principal–agent problem', Wikipedia, the free encyclopedia, http://en.wikipedia.org/w/index.php?title=Principal%E2%80%93agent_problem&printable=, accessed 2 April 2012.

Anon (2010a), 'Rural land disputes lead to unrest in China', *China Daily*, 6 November.

Anon (2010b), 'Officials illegally sell farmer's land', *China Daily Media News*, 4 November.

Bland, B. (2012), 'Vietnam's inconvenient land dispute', *Financial Times* (online), http://blogs.ft.com/beyond-brics/2012/02/13/vietnams-inconvenient-land-dispute/, accessed 27 March 2012.

Chai, J.C.H. (2011), *An Economic History of China*, Cheltenham, UK and Northampton, MA, USA: Edward Elgar.

Clark, C.W. (1957), *The Conditions of Economic Progress*, 3rd edn, London: Macmillan.

Dong, W. (2012), 'Six major problems facing China's "rich economy"', *China Youth Daily*, 14 March.

General Statistical Office of Vietnam (2007). *Results of the 2006 Rural, Agricultural and Fisheries Census*, Hanoi: Statistical Publishing House.

Liu, G. and W. Liang et al. (1987), *China's Economy in 2000*, Beijing: New World Press.

OECD (2010), *OECD Economic Surveys: China*, Paris: Organisation for Economic Co-operation and Development.

Tisdell, C.A. (1993), *Economic Development in the Context of China*, London: Macmillan.

Tisdell, C.A. (2010), 'The survival of small-scale agricultural producers in Asia particularly Vietnam: general issues illustrated by Vietnam's agricultural sector, especially its pig production', in A. Salazaar and I. Riva (eds), *Sustainable Agriculture: Technology, Planning and Management*, New York: Nova Science Publishers, pp. 315–28.

Tisdell, C.A. (2009a), 'Economic growth and transition in Viet Nam and China and its consequences for their agricultural sectors: policy and agricultural adjustment issues', *Regional Development Studies*, **13**, 122–36.

Tisdell, C.A. (2009b), 'Economic reform and openness in China: China's development policies in the last 30 years', *Economic Analysis and Policy*, **39**, 271–94.

Tisdell, C.A. (2009c), *Resource and Environmental Economics: Modern Issues and Applications*, Singapore, Hackensack, NJ and London: World Scientific.

Tisdell, C.A. (2013), *Competition, Diversity and Economic Performance*, Cheltenham, UK and Northampton, MA, USA: Edward Elgar.

Wu, Z. (2009), 'Change in farmland transferring in rural China since 2003', *China Rural Report 2008 of the Research Center for the Rural Economy*, Beijing: Chinese Financial and Economic Press (in Chinese).

Zhao, X. and C.A. Tisdell (2011), 'A comparative economic study of China's and Australia's cotton production: differences, trends and analysis', in T.H. Lee (ed.), *Agricultural Economics, New Research*, New York: Nova Science Publishers, Inc., pp. 37–65.

2. Population, poverty and responsible social protection issues of Asia

Moazzem Hossain

INTRODUCTION

The last three decades are characterized by phenomenal growth in Southeast and East Asian nations.[1] In contrast, South Asia performed moderately over this period. In this chapter an attempt has been made to capture the overall performance of the populous and emerging three regions of Asia from the viewpoint of reducing absolute poverty in prosperous times. More specifically, a comparative analysis of Southeast Asia (including China in the East) and South Asia has been carried out in terms of population growth such as demographic conditions and trends, population distribution, women's participation in the labour force, and demographic challenges since 1990. This analysis contributes partly to answer the frequently asked questions in the literature: Why have the Southeast and East Asian nations done so much better than the countries in South Asia? Is population growth a deterrent to reducing poverty? How can responsible social protection be established for the downtrodden?

This chapter thus focuses on the performance of three regions of Asia (10 nations) in population explosion terms and analyses how far the poverty reduction goals of the Millennium Declaration will be achieved by 2015. The final objective is to examine how to construct responsible social protection measures for the poor, using the case of India.

THE ISSUES

According to the latest estimate, Asia now carries more than half of the total population of the world. Among the most populous countries of the world, ten are in East, Southeast and South Asia: China, India, Indonesia, Pakistan, Bangladesh, the Philippines, Thailand, Viet Nam, Nepal and Sri Lanka. Asia accounts for more than 4 billion people out of the world's 7 billion. More precisely, the figure was 4.4 billion in October 2011 and has been rising (UNFPA 2011).

The Food and Agriculture Organization (FAO) recently estimated that one-sixth of the world population (1020 million) presently suffer from acute hunger, of whom 642 million live in the Asia–Pacific region (FAO 2009). Therefore, East, Southeast, and South Asia are the most critical and alarming regions of the world from the viewpoint of population growth and poverty. By world standards, Asia is characterized by very high rates of population growth, high density and high dependency. Demographically, China is an exception because of its policy of 'a single child only'.

As mentioned earlier, one of the objectives of this chapter is to investigate, given high population growth, the prospects of meeting the poverty target of the Millennium Development Goals (MDGs) in Asia by 2015, the terminal year for achieving the MDGs. In order to investigate this, it is important first to analyse clearly the population and demographic issues of the countries under study. The reason for analysing the demographic dynamics is to inform readers and policymakers about the effect of population growth in the course of reducing poverty. An attempt has been made, first, to outline the general demographic situation in Asia. Next, an analysis on reducing poverty has been attempted, against the backdrop of the population dynamics of the regions.

Poverty eradication is the number one goal of the MDGs. Poverty is both a social and economic evil. Its impact is clearly visible in society in both social and economic terms. The major imbalance in any community is first shown by poverty and its impact on the downtrodden and disadvantaged population of that community. Improvement in all the social issues depends on how poverty changes over a period of time (Quibria 1994).

Before addressing the recent trends in poverty in East, Southeast, and South Asia, it is intended first to clearly define the poverty concept from the viewpoint of the present chapter. Poverty can be conceptualized in terms of food poverty. Food consumption is the most important aspect of human livelihood. It occupies a major share of total consumption of the households in developing countries, particularly in rural areas. It is, therefore, the lack of food (food poverty) which matters most in human welfare in the developing nations including Asian nations.

Food poverty is defined as 'the condition of lacking the resources necessary to acquire a nutritionally adequate diet', and the food poverty line is defined as 'the minimum amount of food an individual must consume to stay healthy' (Green and Thorbecke 1986, p. 60).

The East Asian (from now on East and Southeast Asia combined) nations after more than four decades of strong growth are still unable to solve the problem of food poverty and now a substantial part of their population suffers from food poverty, particularly in China, Indonesia, the Philippines and Vietnam. South Asian populous nations such as India, Pakistan and Bangladesh are in an even worse position with their moderate growth over the last two decades.

A rich and varied literature has grown during the last few decades on the subject of food poverty and the food poverty line (Crawford and Thorbecke 1980; Srinivasan 1981; Sen 1983; Green and Thorbecke 1986). The food poverty line is considered in terms of the nutritional characteristics of the foods, that is, calories (Green and Thorbecke 1986; Hassan and Babu 1991). The energy requirement is composed mainly of two nutrients out of many others: calories and protein. However, Sukhatme (1974) held the view that, given the type of food the human body consumes, the minimum protein requirement will usually be met if calorie requirements are met. The method used with Sukhatme's approach to determine the food poverty line is through the income-calories intake relationship proposed by Srinivasan (1981). Under this method the households can be defined as poor (deficit) that cannot afford to buy their recommended daily needs in calories.

In recent times, however, the World Bank's poverty studies define poverty in terms of population living below US$1 a day per person (PPP $1.25 in 2010) as a minimum poverty line for consistent international comparison, in other words regarded as HCR (head count ratio). This approach uses purchasing power parity (PPP) in measuring the HCR mainly based on consumption of food (World Bank 1996).

Most recently, the UNDP (2010) with the support of several scholars has been projecting poverty as multidimensional, and measurable by a multidimensional poverty index (MPI) (Anand and Sen 1997; Kanbur and Squire 2001; Micklewright and Stewart 2001; Pogge 2009; Alkire and Santos 2010). However, this chapter deals with the income poverty used by the World Bank over the last few decades.

The FAO (2009) suggests that poverty escalated by 100 million people from 2008 to 2009 due to the global financial crisis (GFC). It further reports that although the world has witnessed a significant downturn in people suffering from acute hunger in the 1980s and until the first half of 1990s, in recent years hunger has worsened worldwide. The number of hungry has increased from 825 million in 1995–97 to 873 million in 2004–05, and hit 1.02 billion in 2009.

Out of over a billion global poor, South Asia is the home of more than 600 million. Among these, India alone has more than 350 million poor. Thus, it is important that whatever investigation one attempts on the subject of poverty and poor, it must hear the voices and outcomes from India. The Indian Planning Commission has been at the forefront of investigating poverty in India in terms of defining both poverty and level of poverty that exists in this vast nation.

Dev (2012) finds that the planning commission has begun computing the poverty line for India from the beginning of 1979 based on the Alagh Report. Subsequently, the Lakdawala Committee 1993 and Tendulkar expert group/Committee suggested a new poverty line for India in 2009. The Tendulkar Report moved away from calorie intake as the main indicator for measuring poverty, and instead used price indices of a basket of consumables and services. The methodology used by Tendulkar was basically trying to link the costs of the minimum basket of commodities and income needed to purchase the basket. People whose income is not enough to afford this basket are regarded as poor. Tendulkar drew the line of such an income per capita per month as Rs. 446.68 for 2004–05. These figures are not without controversy.[2] However, in 2011 the poverty line estimate has been more than doubled over 2004–05. The Planning Commission told the Supreme Court in September 2011 that anyone spending Rs. 965 or more per month in urban India and Rs. 781 in rural India would be deemed not to be poor.

While India squabbles about the poverty line, the present study deals with poverty in Asia's populous regions since 1990. Our study concludes that, at the end of the day, Asian poverty has persisted during the last two decades although the regions have been achieving a respectable level of economic growth and in some circumstances the growth rate has been spectacular. Also, it seems that the growth of population over this period, although declining, has not been declining fast enough to achieve desired goals in poverty reduction. Thus, before addressing the question of poverty it is necessary to examine the population dynamics of Asia in more detail.

DEMOGRAPHIC CONDITIONS AND TRENDS IN ASIA

Figure 2.1 presents the UN projections on global population to the year 2100. The UN projection takes into account three possibilities: high, medium and low growth. If one takes the medium rate, bearing in mind the trends in actual (solid line) growth since the 1950s, the world is going to stabilize in population growth in 2080 (middle line). In this respect Asia remains within this time frame and would be likely to stabilize by 2050 (UNFPA 2011).

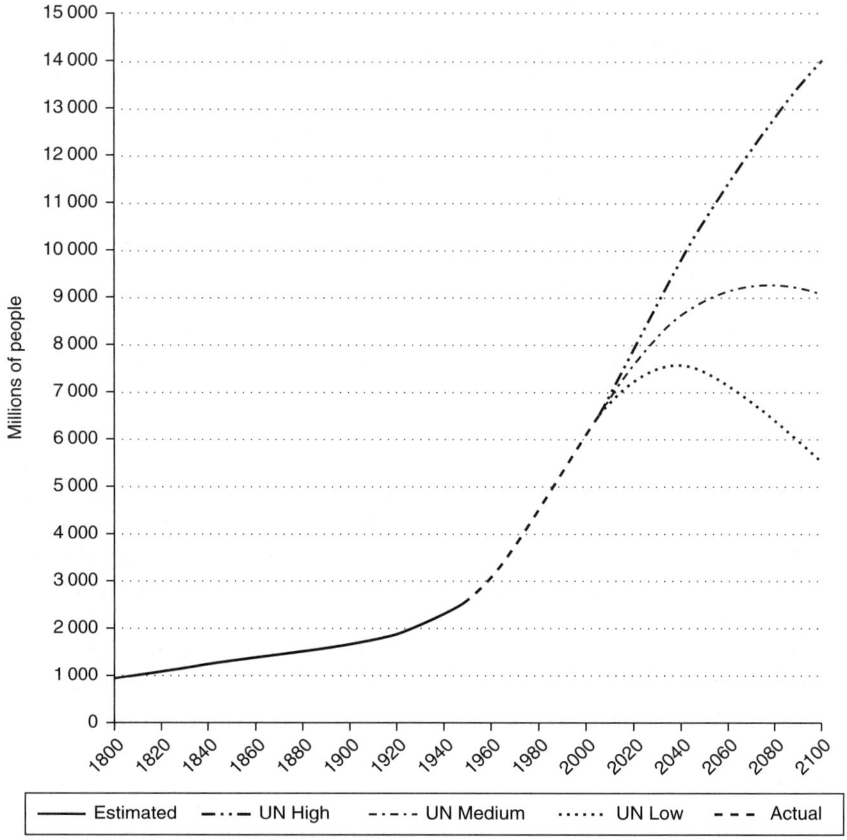

Source: United Nations (2009).

Figure 2.1 UN global projection on population, 1800–2100

In 2012, the Earth Institute at Columbia University, New York, hosted a vigorous debate on global population growth under two headings: 'One planet, too many people?', on 7 March, and 'The population bomb: defused or still ticking?', on 8 March (www.columbia.edu/earthinstitute). The former debated the challenges of planet Earth in terms of food and energy production, supply of water and destruction of environment, given a population projection of 9.5 billion in 2100. The keynote speaker Tim Fox posed two questions, 'Can we manage these needs without depleting our resources and ruining the environment? And how will we do it?' Fox's own answer was, 'There are no insurmountable technical issues'. But barriers to this kind of challenge are 'primarily financial, social and political – no small challenge'.[3] On the projection side of the population, the forum came to the conclusion that, 'if the birth rate rises by half a child per woman, we will wind up with 16 billion people by 2100. If it drops a half, we will end up with six billion – a billion fewer than today'. Thus, the solution seems to lie in increasing voluntary contraception and further access to education, and maternal and early childhood health care.

The second debate, addressed by different experts, came to the same conclusion. A panel of four social scientists from diverse organizations participated, and according to the chair, Earth Institute's Professor Mutter, these panellists shared a total of 121 years' experience in the field of population research. While they bombarded the audience with population numbers and the consequences of an exponential rise in population to 2100, one speaker, Zlotnik, former UN population expert, warned 'while you have been using numbers collected by my former organization, the United Nations, do not trust our numbers . . . You have to look behind them and understand the assumptions that are implied'.

She put forward two observations on combating the crisis on the population front: one, fertility decline does not necessarily mean a lower population or a safer planet. In modern times in high growth nations, people are looking for personal independence. If there are fewer offspring per household, but more people live alone, the number of people in separate households has increased and so has the contribution of more and more greenhouse gas into the atmosphere. Two, she emphasized that longer intervals between births can cause a dramatic drop in the projected population increase. To achieve this, contraceptives and education facilities are needed for the high growth nations.

Bearing all these options in mind, let us see now what has been happening in the 10 populous nations in Asia. Is population going to be a major barrier to realizing the Asian century?

Population Trends and Land Use in Asia

Table 2.1 shows that in 2010 the population of the countries under study is almost half the world's total. Column 6 of the table shows that over 2010–15 the projected growth rate for China alone is very low compared to other nations (0.6 per cent). Among the East Asian nations, the Philippines have the highest rate of growth (1.7 per cent) and Vietnam has the highest density per sq. km (268 people). In South Asia, the highest rate of growth has been witnessed by Pakistan (2.1 per cent) and Bangladesh has the highest density of population (1090 per sq km).

Table 2.2 provides further disaggregated estimates on population and land use in the regions under study. The population of the Philippines

Table 2.1 Demographic conditions

Country	Area (in '000 sq. km)	Average population (million)			Population growth rate		Population ratio (urban/total)	
		1990	2010	2030	1990–95	2010–15	1990	2010**
(1)	(2)	(3)	(4)	(5)	(6)	(7)	(8)	(9)
East Asia	12660	1505	1838	2038	–	–	–	–
China	9597	1142	1354	1463	1.20	0.60	26.4	47.0 (140)
Indonesia	1919	178	233	272	1.50	1.0	30.6	44.3 (122)
Philippines	300	62	94	124	2.3	1.7	48.6	48.9 (240)
Thailand	514	57	68	74	1.2	0.5	29.4	34.0 (126)
Vietnam	330	66	89	105	1.9	1.0	20.3	30.4 (268)
South Asia	4440	1130	1614	2017	–	–	–	–
India	3287	862	1215	1485	2.0	1.3	25.6	30.0 (368)
Pakistan	796	116	185	266	2.4	2.1	30.6	35.9 (202)
Bangladesh	144	116	164	203	2.0	1.3	19.8	28.1 (1090)
Nepal	147	19	30	41	2.5	1.7	8.9	18.6 (190)
Sri Lanka	66	17	20	22	1.1	0.7	18.6	14.3 (304)
Total (East Asia & South Asia)	17100	2635	3452	4055	–	–	–	–

Note: ** Population density is in parentheses in the rightmost column for 2010 projections.

Sources: UNDP (2004); FAOSTAT: World Bank (2004, 2007); UNDP (2004).

Table 2.2 Population estimates and land use

Country	Population density (in per sq. km)*		Land use			
	1995	2005	1990		2005	
			Arable land ha./100 people	Arable land (% of total)	Arable land ha./100 people	Arable land (% of total)
East Asia						
China	126	140	8.1	11.1	8.0	11.1
Indonesia	101	122	10.3	11.2	10.6	12.7
Philippines	230	240	7.4	18.4	7.1	19.1
Thailand	113	126	25.6	34.2	22.4	27.7
Vietnam	201	268	8.2	16.4	8.0	21.3
Averages	154	179	11.92	18.26	11.22	18.38
South Asia						
India	283	368	15.5	54.8	14.8	53.7
Pakistan	163	202	15.2	26.6	14.1	27.6
Bangladesh	833	1090	6.1	70.2	5.7	61.9
Nepal	149	190	9.4	16.0	8.9	16.5
Sri Lanka	273	304	4.7	13.5	4.8	14.2
Averages	340	431	10.18	36.22	9.66	34.78
Asia averages	247	305	11.05	27.24	10.44	26.58

Sources: Human Development Indicators (2004); *FAOSTAT: World Bank (2004, 2007).

and Vietnam has doubled over the 30 years to 2005. China, Thailand and Indonesia's population over the same period has increased by more than one and a half times (World Bank 2007). Land use in the arable category in 2005 was highest in Thailand (27.7 per cent) and lowest in China (11.1 per cent) out of the total surface land of these nations. In South Asia, the population has doubled in almost all the countries under study except Sri Lanka over the last 30 years to 2005 (World Bank 2007). Arable land use in 2005 was highest in Bangladesh (61.9 per cent) and lowest in Sri Lanka (14.2 per cent).

Population Distribution

Distribution of population by age group and dependency ratio by country are presented in Table 2.3. In East Asia in 2005, children in the age

Table 2.3 Distribution of population by age and dependency ratio, 1996–2010

Country	Age group						Dependency ratio	
	1996			2005			1996	2010
	0–14	15–64	65+	0–14	15–64	65+		
(1)	(2)	(3)	(4)	(5)	(6)	(7)	(8)	(9)
East Asia								
China	27.2	66.5	6.3	21.4	71.0	7.6	50.0	39.1
Indonesia	32.2	64.0	3.8	28.3	66.2	5.5	56.0	51.7
Philippines	38.4	58.1	3.5	35.1	61.0	3.9	72.0	60.7
Thailand	26.2	68.2	5.6	23.8	69.1	7.1	47.0	41.2
Vietnam	36.0	58.5	5.5	29.5	65.0	5.4	71.0	45.8
Averages	32	63	5.0	27.6	66.5	6.0	59.2	47.7
South Asia								
India	42.3	60.2	4.0	32.1	62.7	5.3	66.2	55.6
Pakistan	45.5	51.1	3.4	38.3	57.9	3.8	96.6	68.6
Bangladesh	42.3	56.8	0.9	35.5	60.9	3.6	76.2	53.4
Nepal	–	–	–	39.0	57.3	3.7	–	66.6
Sri Lanka	31.7	64.2	4.1	24.1	68.6	7.3	54.5	47.1
Averages	40.5	58.0	3.1	33.8	61.48	4.74	73.4	58.3
Asia averages	36.25	60.5	4.0	12.3	64.0	5.4	66.3	53.0

Sources: US Bureau of the Census, International Data Base, http://www.census.gov/cgi-bin/ipc/idbagg; Hossain et al. (1999); UNDP (2010).

group 0–14 years constitute the second largest proportion of the total population of each country. This is ranged from less than 25 per cent in China and Thailand to more than 29 per cent in Vietnam, Indonesia and the Philippines. The high proportion in Vietnam, Indonesia and the Philippines in this respect reduced the population in the working age group (15–64 years) in 2005. In South Asia, based on 2005 figures the percentage of children below 14 years is highest in Pakistan (38.3 per cent) and Nepal (39 per cent) and lowest in Sri Lanka (24.1 per cent).

Columns 8–9 of Table 2.3 give the dependency ratio as a ratio of children (aged 0–14 years) and older people (65 and above) to persons in the primary working age (15–64 years). The dependency ratios reveal that the Philippines have the highest ratio of dependency in both years in East Asia: 72.0 per cent in 1996, and 60.7 per cent in 2010 among the countries. Also, an important issue shown by columns 8 and 9 is that the dependency ratio has declined for all nations over the 14 year period studied. In South Asia, the dependency ratio has reduced in all the nations over the period

Table 2.4 Female labour force, 1990–2005

Region	Total labour force (millions)		Female labour force (% of total labour force)	
	1990	2005	1990	2005
East Asia				
China	650.0	776.0 (19)	44.8	45.0 (0.2)
Indonesia	75.3	107.0 (42)	38.4	38.0 (−0.4)
Philippines	23.4	37.0 (58)	36.6	40.0 (3.4)
Thailand	30.4	36.0 (18)	46.6	46.0 (0.6)
Vietnam	31.3	44.0 (41)	48.3	49.0 (0.7)
Averages	162.0	200.0 (23.5)	43.0	43.6 (0.6)
South Asia				
India	335.1	435.0 (30)	29.9	28.4 (−1.5)
Pakistan	35.2	56.5 (61)	23.3	27.0 (3.7)
Bangladesh	46.9	63.9 (36)	40.2	36.9 (−3.3)
Nepal	7.1	10.5 (48)	37.9	40.5 (2.6)
Sri Lanka	7.3	8.4 (15)	34.8	30.4 (−4.4)
Averages	86.32	115.0 (33)	33.22	32.64 (−0.58)
Asian averages	124.16	157.5 (27)	38.11	38.12 (.01)

Note: Figures in parentheses show percentage changes over 1990 and 2005.

Source: http://genderstats.worldbank.org.

of 1996 and 2010. The highest dependency is in Pakistan (68.6 per cent in 2010).

In recent years the literature has been flooded with studies on women's participation in developing countries. The UN declaration of 1993 as the Year of Women has contributed to the undertaking of numerous studies by UN agencies on women and their role in future development of developing nations.

Women's participation in a country's labour force is one of the important indicators that measure the role of women in development. Table 2.4 presents detailed comparative information on women's labour force participation for the 10 countries in Asia over 1990 and 2005. Almost half of the Chinese, Vietnamese and Thai labour force was contributed by women in 2005. In Indonesia and the Philippines women contribute less than 40 per cent. In all the countries in East Asia female participation has been steady since 1990 with a moderate average increase of 0.6 per cent for the region. In 2005, among the South Asian nations, the female labour force participation is highest in Nepal (40.5 per cent) and Bangladesh (37 per cent) and lowest in Pakistan (27 per cent). The participation rate in India

Table 2.5 Fertility and contraceptive use rates, 1990–95 and 2010–15

Country	Total fertility rate (birth per woman)		Contraceptive prevalence (% of women aged 15–49)	
	1990–95	2010–15	1995	2005
(1)	(2)	(3)	(4)	(5)
East Asia				
China	2.0	1.8	90	87
Indonesia	2.9	2.0	55	57
Philippines	4.1	2.9	48	69
Thailand	2.1	1.9	72	72
Vietnam	3.3	2.0	75	77
South Asia				
India	3.9	2.5	43	47
Pakistan	5.7	3.6	12	28
Bangladesh	4.0	2.2	46	58
Nepal	4.9	2.9	–	38
Sri Lanka	2.5	2.2	62	70

Sources: http://genderstats.worldbank.org; USAID (2002); UNDP (2010).

is marginally higher than in Pakistan. However, in India, Sri Lanka and Bangladesh, the female participation rate experienced a drop between 1990 and 2005 by about 0.6 per cent.

Asia's Demographic Challenges

By 2010, East Asia's population had reached more than 1.8 billion as against 1.5 billion in 1990. UNDP (2010) projections suggest that by 2030 the total population of the region will reach more than 2 billion. It is however a picture which is far less alarming than in the other parts of Asia, particularly when one looks at population densities. In 2005, Vietnam's population density per sq km reached 268 persons, while China had 140, Indonesia 122, the Philippines 240, and Thailand 126. In contrast, by 2030 South Asia's population will reach 2 billion which is very similar to that of East Asia. In terms of density of population in 2005, while Nepal (190 per sq km), Pakistan (202), Sri Lanka (304) and India (368) have tolerable densities, Bangladesh is certainly in a hugely challenging position (1090).

Table 2.5 presents fertility data for all the nations and suggests that the fertility rate in all these countries has declined over the last ten years. Reduced fertility seems to have helped reduce population in South Asia over the same period. The policymakers have been put in an alarming state that the contraceptive prevalence has not been increasing substantially in the populous nations of the regions. As can be seen from columns 4 and 5 of Table 2.5, between 1995 and 2005 contraceptive prevalence in fact declined from 90 to 87 per cent in China and increased in all other nations, for women aged between 15 and 49, except in Thailand which remained unchanged (72 per cent). This development certainly would not help to reduce the growth rate of population in East Asia to 2030.

South Asia's challenges are even greater in controlling population in the future. The contraceptive prevalence rate is very low in the age group 15 to 49. Only in Sri Lanka is this comparable to East Asian nations. In terms of reduction in fertility rate the South Asia region has a major job to catch up with other parts of Asia. Within South Asia, it appears that in recent years all the nations have been experiencing a decline in initiating projects on population control. It must be emphasized that particularly in Pakistan and Bangladesh the high population growth regime has been bouncing back in recent years after a significant slide in the final quarter of the twentieth century. The increasing trend has been attributed to the lack of budget allocation and decline in support from the international community for direct population control measures. A report by the population and family planning department of the government of Bangladesh identified several issues that are behind the collapse of the population programme implemented over the last two decades. These are: lack of sustained field level workforce, corruption in procurement of relevant effective population control devices and declining morale of the existing workforce due to the lack of initiative by recent governments to keep the dedicated workforce in the department. In particular, it has been reported that the success in the past was achieved by the field level workers who visited the households in rural areas regularly. For example, out of the total rural households at Bangladesh almost 43 per cent were visited in 1993–94. This dropped to only 35 per cent in 1997–98, and in 1999–2000 dropped even further to only 18 per cent of households. The major reason for such a drop in visits was lack of supply of condoms, pills and other population control devices. Moreover, out of more than 50000 field workers, almost 7500 positions are presently vacant (15 per cent of the total). This is no doubt contributing towards destroying the morale of the existing work force and results in the decline of service to the rural people, among whom huge demand for population control devices still exists (Hossain et al. 2010).

The Indian census (2001) reported that the population of this nation rose by 21.34 per cent between 1981 and 2001. Hansen (2002) suggests that in the last several decades, fertility control policies in India have failed to promote a sustainable solution to the problem of population surge. What factors have caused these efforts to disintegrate? The following have been blamed.

India reached the milestone of one billion people in August 1999, which in a country one-third of the United States in size is certainly alarming. India's population more than doubled between 1960 and 2000. In 1960 the population was less than half a billion (431 463 000). It appears that the national sterilization scheme had not achieved a desired outcome, and targeting female sterilization was a failure over the last few decades. In addition, the major challenge for India is the family's strong preference for sons. This has made the overall male and female ratios quite unbalanced in recent years (Begum and Singh 2007). The 2001 census suggests that the number of females per 1000 males was 933, a marginal rise from 927 in the 1991 census (www.indiaeyewitness.com). According to the 2011 census, this figure has dropped, and the number of females per 1000 males is now only 914 (CNN 2012).

As mentioned earlier, Pakistan has the highest rate of population growth in the region. The population control policy has been put on the back burner in Pakistan since the period of 9/11. The United States Agency for International Development (USAID) was the major donor agency in Pakistan, and was the source of funds for the population and family planning department. There was a pause in USAID assistance over 2001 and 2003 as the agency closed its offices in Pakistan immediately after 9/11. USAID was back again in 2008 and extended $60 million to get momentum in population control projects. With the present Taliban insurgency, it is unlikely to put more resources into population control and the problem is going to go from bad to worse in the years to come (*The Nation* 2009).

Sri Lanka was a high achiever in controlling population in the 1970s and 1980s. When the nation went into a civil war in the early 1980s, it almost slipped from the radar of the demographic literature. However, it appears that most recently, the government policy is to satisfy the demand for various family planning methods which has been created by a well-developed and well-promoted fertility control programme during the past phases of policy implementation. Under this policy, greater emphasis is simultaneously being given to improving reproductive health (Dangalle 1998).

ACHIEVEMENTS AND NON-ACHIEVEMENTS IN POVERTY TARGET 2000–08

Having analysed the population issues above in both East and South Asia, let us look at the developments on the MDG front in this part of

Table 2.6 Eradication of extreme poverty

Region	Poverty (PPP $1.25) a day HCR	National poverty line	Share of consumption to poorest quintile (%)
	2000–08	2000–08	2005
East Asia			
China	15.9	2.8	4.7
Indonesia	29.4	16.7	8.4
Philippines	22.6	–	5.4
Thailand	<2	–	6.3
Vietnam	21.5	28.9	9.0
South Asia			
Bangladesh	49.6	40.0	9.0
India	41.6	28.6	8.9
Nepal	55.1	30.9	6.0
Pakistan	22.6	–	9.3
Sri Lanka	14.0	22.7	8.3

Sources: Hossain (2007); World Bank (2007); UNDP (2010).

the world. All the multilateral agencies including the IMF and the World Bank are fully committed to implementing the eight human development and poverty alleviation goals by 2015.[4] Given the conditions in population growth in both East and South Asia, an attempt has been made here to investigate how the countries under study are coping with their commitments to achieve goal 1 (reduction of poverty by half from 1990 level) of the MDGs by 2015. The investigation has been carried out by analysing the achievements between 2000 and 2008, the first half of implementing the MDG strategy.

In terms of eradicating extreme poverty certainly East Asian nations have been ahead of South Asia (Table 2.6). The headcount ratio suggests that half of the population in Bangladesh and 41 per cent in India live on or under PPP $1.25 a day per capita while a lower level of poverty in PPP terms exists in the rest of the nations between 2000 and 2008 (Table 2.6). Nepal's position is even worse than that of Bangladesh and India, at 55 per cent. It is unlikely that these nations are going to reach the MDG target of halving poverty by 2015. In East Asia, Indonesia has almost 30 per cent of its population under the poverty line, while the Philippines and Vietnam are at a significantly lower level. Given China's prosperity over the last three decades, its poverty level seems remained unusually high (almost 16 per cent of the total, Table 2.6).

Taking share of consumption by the poor into consideration, China,

Table 2.7 Results of econometric analysis

Cross section observation	10 nations under study	
Intercept	−26.2436	(−2.28)
Population growth	11.3209	(3.24)
Per cent of rural population	0.3527	(2.3)
Population density	0.0186	(2.27)
R value	0.65	
Adjusted R	0.58	

Note: Figures in parentheses are t-values.

Source: Hossain and Selvanathan (2011).

the Philippines and Thailand's poor have the lowest consumption share compared to Indonesia and Vietnam among East Asian nations. In South Asia, Nepal's poor have the lowest consumption share. This certainly needs to be improved. It is disturbing to see from the analysis above that the populous nations of East and South Asia, in population density terms (China, Indonesia, the Philippines, Bangladesh and India) have been making slow progress in eradicating absolute poverty. In China and Indonesia respectively 16 and 30 per cent of the population lives in poverty, and in Bangladesh and India respectively more than 49 and 41 per cent live in poverty (Table 2.6 in HCR terms).

In the rest of this section the impact of population on poverty reduction has been illustrated in simple econometric terms. Hossain and Selvanathan (2011) employed a regression model with percentage of population below poverty line (BPL) as the dependent variable, and population growth, percentage of rural population and population density per square kilometre as independent variables.

Table 2.7 provides estimated results from this study. As can be seen, all the estimated coefficients are statistically significant at the 5 per cent level. The parameter estimates show that population growth has the right sign and is significantly associated with poverty. Rural population and population density also picked up positive co-efficients and are statistically significant.

The key results are:

- The coefficient on population growth is both positive and significant at 5 per cent level of significance. The result suggests that with a 0.5 per cent growth in population, poverty will increase by more than 5.5 per cent, quite a large impact.

- The positive and highly significant rural population estimates suggest that with a 10 per cent increase in rural population, poverty will increase by more than 3.5 per cent.
- The coefficient on population density is both positive and significant and its estimated value is only 0.0186. This implies that a 10 per cent increase in population density, on average, will result in only 0.19 per cent increase in poverty. This is certainly a negligible impact.

In summary, this analysis with cross sectional data confirms that population growth has a strong impact on the increase of poverty which is consistent with the generic view in Asia. The impact of rural population ratio on poverty is found to be moderate however, once again consistent with the generic view. The population density has a low impact on poverty, which suggests that the population density is not yet critical in the populous regions of Asia. In this regard, there seems room for improvement in populating those areas which are still seen to be less inhabited.

BUILDING RESPONSIBLE SOCIAL PROTECTION FOR THE POOR

It is amply clear from the above analysis that at the end of the first decade of the twenty-first century, an unacceptable level of poverty exists in Asia, although the continent has been enjoying prosperity over the last three decades with a record level of growth never seen before. Indeed, out of the total poor, almost half lives in South Asia and one-quarter lives in the East. How will it be possible to realize the Asian century in the midst of such a huge population under hunger and poverty? It is thus essential that the region must make new strategies and bring into action some responsible policy measures to protect almost one-third of its population (out of a total 4 billion) from hunger, to make Asian prosperity meaningful. In this final section, an attempt is made to analyse a pro-poor strategy which is currently being adopted by the present government of India. Why India? This nation has more than 350 million people poor (almost half of Asia's poor). As the world's largest democracy, which entered into its seventh decade of independence in 2007, India cannot wait any more to solve its poverty problem. Sixty-five years is a very long time. Most importantly, if India can set an example in making a pro-poor strategy, other smaller nations of Asia will be in a position to emulate it effectively.

Table 2.8 Growth rates of South Asian nations, 1961–2007

	GDP growth (%)					
	1961–70	1971–80	1981–90	1991–2000	2000–07	2012*
Bangladesh	4.1	1.6	3.9	4.8	6.3 ($2000)	6.0
India	3.9	3.1	5.9	5.3	7.0 ($3800)	7.2
Pakistan	7.3	4.6	4.3	5.2	5.5 ($2800)	5.0
Sri Lanka	4.6	4.3	4.4	5.2	4.3 ($5000)	6.0

Notes:
* Estimates.
Figures in parentheses are per capita income in PPP $ for 2005.
World average is $9000.

Sources: ADB (2004, 2008); World Bank (various).

South Asia's Poor: Who and Where?

The region has more than half a billion poor, and among them at least 400 million are regarded as hard core. The hard core poor are those who have no access to two square meals a day, no shelter, no health facility and no proper clothing. Education facilities for poor children are limited. It is not the percentage of the total population that matters most, it is the absolute number, living in urban slums and rural huts. Indeed, 400 million is a very large number by twenty-first century standards. Slums in all big cities of South Asia have been flooded with new arrivals every day, but the rural poor need close and special attention since even now almost 70 per cent live in a rural environment.

To deal with this, one needs to present first the relative prosperity South Asia has enjoyed over the last two to three decades in GDP terms. Table 2.8 presents the impressive economic growth enjoyed by the region before the GFC began in 2008. India's sustained growth of 7 per cent per annum over almost a decade has brought major optimism about emerging India both inside and outside. A study suggests that a high growth rate makes a huge difference to employment in the region. Islam (2012) finds, 'On the side of labour, every percentage point of GDP growth has been found to create some 250000 jobs'. Hence, India has created more than half a million jobs every year over the last decade. Is that enough to alleviate poverty in India in a reasonable time frame? The recent poverty estimates by the Planning Commission's Tendulkar Committee found poverty (national measure) declined by 7.3 per cent from 37.2 per cent in 2004–05 to 29.8 per cent in 2009–10 (see Appendix 2.1).

It appears that India's decline in poverty has been in line with the prosperity the nation has been enjoying, with a sustained growth in GDP over almost a decade. However, the Planning Commission's claim of rapidly declining poverty was not taken lightly by various quarters in India. Particularly, the opposition in parliament (BJP) threw out the report, saying that it took the level of ultra poor as the poverty line (Desher Sakal 2012; *Times of India* 2011). The government has taken the opposition's view seriously and formed a commission to examine the methods and the validity of the claim of declining poverty endorsed by the Planning Commission (GoI 2012). It remains to be seen how the enquiry unfolds.

Who are the poor?

Appendix 2.1 further identifies the poor by social and religious groups:

Poverty ratio by social groups:

- In rural areas, Scheduled Tribes (STs) exhibit the highest level of poverty (47.4 per cent), followed by Scheduled Castes (SCs) (42.3 per cent), and Other Backward Castes (OBCs) (31.9 per cent), against 33.8 per cent for all classes.
- In urban areas, SCs have HCR of 34.1 per cent followed by STs (30.4 per cent) and OBCs (24.3 per cent) against 20.9 per cent for all classes.
- In rural Bihar and Chhattisgarh, nearly two-thirds of SCs and STs are poor, whereas in states such as Manipur, Orissa and Uttar Pradesh the poverty ratio for these groups is more than half.

Poverty ratio by religious groups:

- Sikhs have the lowest HCR in rural areas (11.9 per cent), whereas in urban areas Christians have the lowest proportion (12.9 per cent) of poor.
- In rural areas, the HCR for Muslims is very high in states such as Assam (53.6 per cent), Uttar Pradesh (44.4 per cent), West Bengal (34.4 per cent) and Gujarat (31.4 per cent).
- In urban areas the poverty ratio at all India level is highest for Muslims (33.9 per cent). Similarly, for urban areas the poverty ratio is high for Muslims in states such as Rajasthan (29.5 per cent), Uttar Pradesh (49.5 per cent), Gujarat (42.4 per cent), Bihar (56.5 per cent) and West Bengal (34.9 per cent).

Where are the poor?

Whatever the outcome of the enquiry commission of the government, it is indeed interesting to note where the major portion of poor people in India live. The Planning Commission provides some indications in this regard. Out of 35 states, almost 10 populous states have been the major sources of poverty. In three of these (Bihar, Chattisgarh and Manipur), almost half of the people are poor. In absolute terms, the number of poor people in these states is more than 67 million. In the remaining seven states, poverty exists among one-third of the population. In absolute terms, the number of people who are poor is 140 million in these states (Assam, Jharkhand, Madhya Pradesh, Orissa, Uttar Pradesh, Dadra and Nagar Haveli, and Daman and Diu). The 10 states combined have 200 million poor out of India's 355 million in all 35 states of the nation (GoI 2012, Table 3).

India's Food Security Bill: A Step in the Right Direction?

With such huge poverty in existence historically, efforts were made to put very strong social safety net measures in place in India, immediately after World War II. The colonial government saw prices rising during the war and established measures such as price controls and a food rationing system. Unfortunately, these measures did not work and as a consequence the great Bengal famine took place in 1943, when millions of people perished. After Independence in 1947, the Indian Planning Commission took charge of making economic plans and preparing development strategy, with the first five year plan in the early 1950s, largely following the import substitution (IS) approach. This continued until the economic reform process began in the early 1990s. Since a huge number of people remained in poverty in both rural and urban India, the successive governments continued various social safety net measures following a dual policy system: market driven prices of food grains and government controlled prices.[5]

One of the most popular schemes, introduced in 2008 nationwide (596 districts), is The Indian Mahatma Gandhi National Rural Employment Guarantee Scheme (NREGS), regarded as the world's largest public employment programme 'aimed at enhancing the livelihood security of rural inhabitants' (Sharma and Gulati 2012). This scheme claims that about 250 million people have been hired under the project, with a cost to government coffers of 0.51 per cent of GDP and about 3.3 per cent of budget expenditure each year. The beneficiaries generally come from the most marginalized groups such as CSs and STs. The scheme guarantees a statutory right to rural households to claim up to 100 days of wage-employment in a fiscal year.

In late 2011, the present government proposed an even larger scheme to protect the livelihoods of the Indian poor by introducing the 'Food Security Bill' in parliament.

Table 2.9 Indian Food Security Bill

What does it offer?	Legal entitlement to cheap grain and millets
Who will get it?	63.5% of the total population 75% in rural and 50% in urban India 46% of rural coverage will be for priority households (HH) (below poverty line) 28% of the urban coverage will be for priority HH The rest will be general (above poverty line) HH
How much?	7kg per person or 35kg per family per month for priority HH 3kg per person per month for general households
At what price?	For priority group, millets Rs. 1/kg, wheat Rs. 2/kg, rice Rs. 3/kg Lower than present ration rates of Rs. 3, Rs 4.15 and Rs. 5.65 each General HHs will have to pay 50% of the minimum price
What about source of supply?	Production has to be increased from presently 54 million tonnes to 62 million tonnes

Source: Newspaper reports (several, December 2011).

The main aim of the programme is to provide for 'food and nutritional security . . . by ensuring access to adequate quantity of quality food at affordable prices, for people to live a life with dignity'. The bill is bigger than that for the rural job guarantee scheme (NREGS) described above. According to government estimates, this law will cost the treasury at least $1.9 billion, which is $633 million above the current subsidy provided to the poor in various food rationing programmes.

This programme is regarded as the greatest experiment in the world in providing food security to the millions of poor (see Table 2.9).

In this regard, it is appropriate to put forward the views of the wider world on this initiative. Some world famous economists of Indian origin wrote an opinion piece in the *Hindustan Times* on 29 February 2012 insisting that 'It simply can't fail'. They put forward in this public forum two simple amendments for successful implementation of the scheme. These are:

> First, the proposed framework based on three categories of households – priority, general and excluded – is doomed to create the same sort of problems as below poverty line (BPL) targeting . . . Why not, therefore, just have two categories: the included and the excluded, and treat all included groups equally? . . .

> Second, we believe that the Bill should allow greater flexibility to state governments as long as they stay within the resource limits specified by the central government. Each state must be left free to carry out the public distribution system (PDS) reform as its sees fit. Some states may want to change to cash transfers, others may opt to continue with transfers in kind, or go to something intermediate, such as food stamps . . .
>
> This bill will unveil the largest poverty alleviation programme in India. It's the most progressive piece of legislation ever contemplated in India. We simply cannot let it fail.

- Abhijit V. Banerjee is Professor, Massachusetts Institute of Technology
- Pranab Bardhan is Professor, University of California, Berkeley
- Ashok Kotwal is Professor, University of British Columbia
- Milind Murugkar is Policy Analyst, Pragati Abhiyan, Nasik
- Bharat Ramaswami is Professor, Indian Statistical Institute, New Delhi

By all means, compared to present and past measures in the area of social protection, the Food Security Bill of India is a huge step in the right direction to protect India's millions of poor from hunger in the years and decades to come. This is a true pro-poor strategy which has broken all the records of the past in India and elsewhere. If implemented as suggested above, let this be the first step in realizing the Asian century.

CONCLUSIONS

Among the socio-economic targets of the MDGs, it is the poverty eradication which appears to be lagging behind compared to the expectations at the time of the Millennium Declaration in 2000. There are many reasons, but the above analyses suggest that the continuing pressure from population remains one of the major hurdles to alleviating poverty. The populous nations of these regions, particularly those of South Asia, have been falling further behind in achieving the poverty reduction goal of the MDGs.

China, Indonesia and the Philippines have in the range of 16 to 30 per cent of their population in acute poverty (defined as less than PPP $1.25 a day per person). In absolute terms, this means that more than 200 million people remain poor. In the case of South Asia it is even worse. The South Asian populous nations (India and Bangladesh) carry a burden in the range of 41 to 50 per cent respectively of their people living in acute poverty. In absolute terms this is more than 400 million in these two nations alone.

The outcome of a cross sectional regression study for ten nations in Asia suggests that for a half of 1 per cent growth in population, the poverty will increase by 5.5 per cent, a huge impact indeed. Thus, the present study takes the view that the current growth trend in population, crisis in income distribution and the GFC combined will make matters worse in these regions in the days and years to come and they certainly will not be in a position to attain the targets stipulated in the Millennium Declaration in 2000. The MDG 1 will drag further and further behind; thus it is impossible to predict achievements until the global economy recovers from the current financial meltdown and the responsible social protection measures are in place.

Having said this, a promising strategy has been unfolding in the Indian parliament with the passing of legislation on food security. This legislation, if implemented soon in its entirety, will be a landmark step for the Indian government, to show responsibility in making poverty history in this part of the world. Of course, this is a very ambitious plan to protect the unprotected (poor) of society. The Indian economy without doubt can afford it in the current time of prosperity, with more than 7 per cent sustained growth in the past decade, projected to be even better in the future. Let India show the world a right direction in unveiling a meaningful pro-poor strategy, which would certainly make the twenty-first century the much talked-about century for Asia.

NOTES

1. See worldbank.org or adb.org for regional definition.
2. See *Times of India*, 21 September 2011.
3. See more in Fox (2011).
4. See poverty reduction strategy papers (PRSPs) developed by all developing nations in www.imf.org
5. See Kumar (2010) for other targeted distribution programmes.

REFERENCES

ADB (2004), *Asian Development Bank Outlook 2004*, Manila: Asian Development Bank.

ADB (2008), *Asian Development Bank Outlook 2008*, Manila: Asian Development Bank.

Alkire, S. and M. Santos (2010), 'Acute multidimensional poverty: a new index for developing countries', *Human Development Research Paper 11*, New York: UNDP.

Anand, S. and A. Sen (1997), 'Concepts of human development and poverty: a multidimensional perspective', *Human Development Report 1997*, New York: UNDP.

Banerjee, A.V., P. Bardhan, A. Kotwal, M. Murugkar and B. Ramaswami (2012), 'It simply can't fail', *Hindustan Times*, Kolkata, English daily, 29 February.

Begum, S.M. and C. Singh (2007), 'The economics behind son preference in South India', in Allam Ahmed (ed.), *World Sustainable Development Outlook 2007*, Brighton: University of Sussex, pp. 391–400.

CNN (2012), News Report, Amanpour, New York, 11 June.

Crawford, E. and E. Thorbecke (1980), 'The analysis of food poverty: an illustration from Kenya', *Pakistan Development Review*, **19** (4), 311–35.

Dangalle, N. (1998), 'Controlling population growth in Sri Lanka: perceptions, policies and strategies', *Journal of Population Studies*, **1** (1), 1–24.

Desher Sakal (2012), *Bengali Daily*, Kolkata, 21 March.

Dev, M.S. (2012), 'Poverty line: usefulness of poverty data', *The Economic Times*, 28 March.

FAO (2009), *The State of Food Insecurity in the World (SOFI): The 2009 Hunger Report*, Rome: Food and Agriculture Organization.

Fox, T. (2011), 'One planet, too many people?', Report produced for the Institute for Mechanical Engineers, London: IME.

GoI (2012), 'Press note on poverty estimates, 2009–10', New Delhi: Planning Commission, March.

Green, J. and E. Thorbecke (1986), 'A methodology for measuring food poverty: applied to Kenya', *Journal of Development Economics*, **24** (1), 59–74.

Hansen, Z. (2002), *Population Control in India: A Review*, Waterville ME: Colby College.

Hassan, R.M. and S.C. Babu (1991), 'Measurement and determinants of rural poverty: household consumption patterns and food poverty in rural Sudan', *Food Policy*, **6** (6), 451–60.

Hossain, M. (2007), 'Human development and poverty alleviation in South East and East Asia: a post-crisis account', in C. Tisdell (ed.), *Poverty, Poverty Alleviation and Social Disadvantage*, New Delhi: Serials Publications, pp. 750–62.

Hossain M. and E. Selvanathan (2011), 'Population, poverty and CO2 emissions in Asia: an overview', in M. Hossain (ed.), *Climate Change and Growth in Asia*, Cheltenham, UK, Northampton, MA, USA: Edward Elgar, pp. 17–37.

Hossain M., Islam I. and R. Kibria (1999), *South Asian Economic Development: Transformation, Opportunities and Challenges*, London and New York: Routledge.

Hossain M., Kathuria, R. and I. Islam (2010), *South Asian Economic Development: 2nd edition*, London and New York: Routledge.

Human Development Indicators (2004), http://www.infoplease.com/ipa/A0004379.html

Islam R. (2012), 'Macroeconomic policy for inclusive development', *The Daily Star*, Dhaka, English daily, 16 March.

Kanbur, R. and Squire, L. (2001), 'The evolution of thinking about poverty: exploring the interactions', in G. Miere and J.E. Stiglitz (eds), *Frontiers of Development Economics: The Future Perspective*, New York: Oxford University Press, pp. 183–226.

Kumar, P. (2010), *Targeted Public Distribution System: Performance and Inefficiencies*, New Delhi: Academic Foundation and National Council for Applied Economic Research.

Micklewright, J. and K. Stewart (2001), 'Poverty and social exclusion in Europe: European comparisons and impacts on enlargement', *New Economy*, **8** (2), 104–9.
Myrdal, G. (1968), *Asian Drama: An Inquiry into the Poverty of Nations*, New York: Pantheon Books.
Myrdal, G. and A. Myrdal (1970), 'The challenge of world poverty: a world anti-poverty program in outline', *The Christian A. Herter Lecture Series*, New York: Pantheon Books.
Pogge, T. (2009), *Developing Morally Plausible Indices of Poverty and Gender Equity: A Research Program*, New York: New York University Institute of Public Knowledge.
Quibria, M.G. (ed.) (1994), *Rural Poverty in Developing Asia I, Bangladesh, India, Sri Lanka*, Manila: Asian Development Bank.
Sen, A.K. (1983), 'Poor, relatively speaking', *Oxford Economic Papers*, **35** (2), 219–31.
Sharma, P. and A. Gulati (2012), *Approaches to Food Security in Brazil, China, India, Malaysia, Mexico, and Nigeria: Lessons for Developing Countries*, New Delhi: ICRIER Policy Series, no. 14.
Sukhatme, P.V. (1974), 'The protein problem: its size and nature', *Journal of the Royal Statistical Society*, Series A, **137**.
Srinivasan, T.N. (1981), 'Malnutrition: some measurement and policy issues', *Journal of Development Economics*, **8**, 3–19.
The Nation (2009), '$60 million population control programme launched', Karachi, English daily.
Times of India (2011), News Report, New Delhi, 21 September.
UN (2009), *UN World Population Projections to 2100*, New York: United Nations.
UNDP (2004), *Human Development Report*, Oxford: Oxford University Press.
UNDP (2010), *Human Development Report*, New York: Palgrave Macmillan.
UNFPA (2011), *World Population Report 2011*, New York: United Nations Funds for Population.
USAID (2002), *Global Population Profile.*
World Bank (1996), *World Development Report 1996, From Plan to Market*, Oxford: Oxford University Press.
World Bank (2004), *World Development Indicators*, Washington, DC: The World Bank.
World Bank (2007), *World Development Indicators*, Washington, DC: The World Bank.

Websites

www.censusu.gov/cgi-bin/ipc/idbagg
http://www.census.gov/cgi-bin/ipc/idbagg
www.columbia.edu/earthinstitute
www.devdata.worldbank.org (2005).
www.devdata.worldbank.org (2004).
www.devdata.worldbank.org (2001).
www.devdata.worldbank.org (2000).
www.fao.org
www.genderstats.worldbank.org

www.indiaeyewitness.com
www.infplease.com/ipa/A0004379.html
http://www.infoplease.com/ipa/A0004379.html
www.unfpa.org

APPENDIX 2.1

Government of India, Planning Commission: Press Note on Poverty Estimates, 2009–10, New Delhi, 19 March 2012

The Tendulkar Committee for the first time recommended use of implicit prices derived from quantity and value data collected in household consumer expenditure surveys for computing and updating the poverty lines. Tendulkar Committee developed a methodology using implicit prices for estimating state wise poverty lines for the year 2004–05. Using these poverty lines and distribution of monthly per capita consumption expenditure based on mixed reference period (MRP), the Tendulkar Committee estimated poverty ratios for the year 2004–05. In its Report, Tendulkar Committee recommended a methodology for updating 2004–05 poverty lines derived by it.

2. Accordingly, implicit price indices (Fisher Price Index) have been computed from the 66th Round NSS (2009–10) data on Household Consumer Expenditure Survey. As per Tendulkar Committee recommendations, the state wise urban poverty lines of 2004–05 are updated for 2009–10 based on price rise during this period using Fisher price indices. The state wise rural-urban price differential in 2009–10 has been applied on state specific urban poverty lines to get state specific rural poverty lines.

3. The HCR is obtained using urban and rural poverty lines which are applied on the monthly per capita expenditure (MPCE) distribution of the states. The aggregated BPL population of the states is used to obtain the final all-India HCR and poverty lines in rural and urban areas. Some of the key results are:

   - The all-India HCR has declined by 7.3 percentage points from 37.2 per cent in 2004–05 to 29.8 per cent in 2009–10, with rural poverty declining by 8.0 percentage points from 41.8 per cent to 33.8 per cent and urban poverty declining by 4.8 percentage points from 25.7 per cent to 20.9 per cent.
   - Poverty ratio in Himachal Pradesh, Madhya Pradesh, Maharashtra, Orissa, Sikkim, Tamil Nadu, Karnataka and Uttarakhand has declined by about 10 percentage points and more.
   - In Assam, Meghalaya, Manipur, Mizoram and Nagaland, poverty in 2009–10 has increased.
   - Some of the bigger states such as Bihar, Chhattisgarh and Uttar Pradesh have shown only marginal decline in poverty ratio, particularly in rural areas.

Poverty ratio for social groups

- In rural areas, STs exhibit the highest level of poverty (47.4 per cent), followed by SCs (42.3 per cent), and OBCs (31.9 per cent), against 33.8 per cent for all classes.
- In urban areas, SCs have HCR of 34.1 per cent followed by STs (30.4 per cent) and OBCs (24.3 per cent) against 20.9 per cent for all classes.
- In rural Bihar and Chhattisgarh, nearly two-thirds of SCs and STs are poor, whereas in states such as Manipur, Orissa and Uttar Pradesh the poverty ratio for these groups is more than half.

Among religious groups

- Sikhs have lowest HCR in rural areas (11.9 per cent) whereas in urban areas, Christians have the lowest proportion (12.9 per cent) of poor.
- In rural areas, the HCR for Muslims is very high in states such as Assam (53.6 per cent), Uttar Pradesh (44.4 per cent), West Bengal (34.4 per cent) and Gujarat (31.4 per cent).
- In urban areas the poverty ratio at all India level is highest for Muslims (33.9 per cent). Similarly, for urban areas the poverty ratio is high for Muslims in states such as Rajasthan (29.5 per cent), Uttar Pradesh (49.5 per cent), Gujarat (42.4 per cent), Bihar (56.5 per cent) and West Bengal (34.9 per cent).

For occupational categories

- Nearly 50 per cent of agricultural labourers and 40 per cent of other labourers are below the poverty line in rural areas, whereas in urban areas, the poverty ratio for casual labourers is 47.1 per cent.
- As expected, those in regular wage/salaried employment have the lowest proportion of poor. In the agriculturally prosperous state of Haryana, 55.9 per cent of agricultural labourers are poor, whereas in Punjab it is 35.6 per cent.
- The HCR of casual labourers in urban areas is very high in Bihar (86 per cent), Assam (89 per cent), Orissa (58.8 per cent), Punjab (56.3 per cent), Uttar Pradesh (67.6 per cent) and West Bengal (53.7 per cent).

Based on the education level of head of the household

- In rural areas, as expected, households with 'primary level and lower' education have the highest poverty ratio, whereas the reverse is true for households with 'secondary and higher' education. Nearly two-thirds of households with 'primary level and lower' education in rural areas of Bihar and Chhattisgarh are poor, whereas it is 46.8 per cent for UP and 47.5 per cent for Orissa.
- The trend is similar in urban areas.

For categories by age and sex of head of the household

- In rural areas, it is seen that households headed by minors have poverty ratio of 16.7 per cent and households headed by female and senior citizens have poverty ratio of 29.4 per cent and 30.3 per cent respectively.
- In urban areas, households headed by minors have poverty ratio of 15.7 per cent and households headed by female and senior citizen have poverty ratio of 22.1 per cent and 20.0 per cent respectively against overall poverty ratio of 20.9 per cent.

4. State wise details of poverty lines for 2009–10, poverty ratios for 2009–10 and poverty ratios for 2004–05 are given in Table 1, Table 2 and Table 3 respectively (see GoI 2012).

3. Resource constraints and Asia's growth: regional cooperation for enhancing energy security

Shanawez Hossain and Tapan Sarker

INTRODUCTION

Based on the recent historic trend of economic growth and the prospect of further growth in coming decades, profound economic transformation is expected with the realization of the 'Asian century' by around 2050, with Asia sharing 51 per cent of world gross domestic product (GDP) (at market exchange rates) (Kohli et al. 2011). However, the prospect of the Asian century depends heavily on the ability of Asian countries to navigate a resource constrained social, political and economic context. Developing[1] and emerging[2] Asia is more vulnerable to these constraints due to higher projected risks associated with climate change and lower adaptation capacity. By 2030 developing countries will require US$28–67 billion or 0.2–0.8 per cent of global investment flows to enable adaptation to climate change (UNFCC 2007). Resource constraint issues, both knowledge and financial, remain the most important consideration for Asia to realize the Asian century, with serious implications for growth, as well as other factors including the environment, climate, peace, prosperity and harmony. Despite Asia's remarkable economic growth, the region's water, food and energy security is not ensured, as a wide disparity still exists across regions and between and within countries. These interrelated aspects are critical for Asia's long-term growth and development because the future competitiveness and prosperity of Asian nations will largely depend on their level of efficiency in the use of natural resources and on their progress in the low carbon race (Kohli et al. 2011).

This chapter analyses the water–food–energy nexus, mainly from the perspective of developing Asia. The important question that this chapter addresses is whether resource constraints will disrupt Asia's growth, and adversely affect its stability, peace, harmony and prosperity in relation to realizing the Asian century. It also explores prospects for cooperation at the national, regional and sub-regional levels to utilize resources more effectively, as well as taking 'Resource Centered Strategies (RCS)' to overcome constraints related to law, policy and financing. Considering the high significance of energy among these three resources, this chapter examines issues related to the energy sector more elaborately. Taking financing into consideration as the major constraint, it explores how domestic and international funds can best be used to ensure energy security in Asia. It examines the role that Asian regional institutions can play in addressing financial and policy constraints at national, sub-regional and regional levels. Using a sustainable enterprise economy perspective, the chapter explores what can be done at national and regional levels to promote eco-innovation and looks at what regional cooperative framework should be explored to promote green growth in Asia.

The chapter is organized as follows. The following section examines the nexus between water, food and energy resources from developing Asia's perspective in order to understand present and future security trends. Then the energy sector will be analysed in more detail to provide an understanding of the challenges and potential for this sector, considering a 2050 timeline as well as a mid-term timeline of 2030. In particular, the following issues will be explored in detail: issues in relation to policy gaps, financing needs, financing instruments, and the role of public–private, regional and international institutions. Examples of the best practices from different parts of Asia will be cited in order to identify possibilities to replicate these practices elsewhere. Benefits of potential regional cooperation and integration mechanisms will also be examined in detail to propose a policy framework. The final section proposes a framework for cooperation at national, sub-regional and regional levels to address energy security related challenges in realizing the Asian century.

RESOURCE CONSTRAINTS AND GROWTH NEXUS IN ASIA: AN OVERVIEW

Asia, a land of around 3.96 billion people in 2010, is expected to consist of around 3.77 billion or 53 per cent of the global population in 2050 (UN 2011). Similarly, projected economic growth could result in a per capita GDP of $38 600 (purchasing power parity (PPP)) in Asia in 2050, similar to Europe today (Kohli et al. 2011). More significantly, projections indicate that Asia's future growth will be led mainly by developing countries in Asia. Table 3.1 shows that in future decades, the GDP growth rate of China, Association of Southeast Asian Nations (ASEAN), India and other South Asian countries will surpass many of the current developed

Table 3.1 Growth projection of Asia

Region	2010			2030			2010–30
	Population	GDP/ capita	World share	Population	GDP/ capita	World share	GDP growth
ASEAN	593.4	2 639	2.64	706.0	6 564	3.86	5.6
China	1 348.9	4 233	9.63	1 402.3	15 748	18.39	7.0
South Asia							
India	1 224.6	1 229	2.54	1 523.5	4 312	5.47	7.6
Bangladesh	148.69	650	0.32	181.86	2 373	0.36	7.8
Bhutan	0.72	1 903	0.01	0.89	4 995	0.004	6.2
Maldives	0.31	4 403	0.00	0.38	10 530	0.003	5.6
Nepal	29.95	455	0.05	39.94	4 349	0.06	8.6
Pakistan	173.59	1 005	0.61	234.43	3 690	0.72	8.3
Sri Lanka	20.85	2 126	0.14	23.09	5 286	0.10	5.2
NIEs	71.2	17 785	2.13	73.3	33 539	2.05	3.4
Japan	126.5	40 444	8.63	120.2	52 749	5.28	1.1
USA	310.4	46 494	24.32	361.7	62 961	18.97	2.3
Europe	511.3	34 491	29.72	527.8	49 252	21.65	2.0
World	6 641.1	8 933	100.00	7 932.8	15 135	100.00	3.6

Notes:
Population: millions.
GDP/capita: 2009 US$.
NIEs = Newly industrialized economies in Asia, including: Republic of Korea, Hong Kong, China, Singapore, Taipei and China.

Sources: World Bank (2011); Centennial Group Holdings (2011); authors' own calculation.

countries of Asia and the world. China in particular, with an average growth rate of around 7 per cent during the period 2010–30, will increase its total world share of GDP from around 10 per cent in 2010 to around 18 per cent in 2030. India's share of world GDP will also double in this time, from 2.54 per cent in 2010 to 5.47 per cent in 2030. Similarly, many other developing countries will also be making significant contributions to Asia's projected growth in the next decades.

As of 2009, Asia is home to about 1.8 billion people living on less than $2 per day: about 1.5 billion have no access to decent sanitation, 1.2 billion people lack access to roads, 930 million have no access to electricity, and 638 million have no access to clean water (ADB 2009). Therefore, it is clear that despite the last four or five decades' rapid economic growth of Asia, many Asian economies still failed to improve the quality of life (QoL) for the majority of their people, and considerable disparities exist among countries and within countries, through both the income and non-income dimensions of life. For instance, in 1990, only 72 per cent of the population had access to improved water facilities as compared to 18 per cent who had access to improved sanitation. However, the situation has improved slightly in 2008 with 88 per cent and 31 per cent of the population having access to improved water and sanitation facilities respectively (World Bank 2011). A number of Asian countries have made significant improvement in access to improved water facilities and sanitation in recent years. For example, in China only 67 per cent and 41 per cent of the population had access to improved water facilities and sanitation in 1990 as compared to 89 per cent and 55 per cent in 2008. Similarly, in Cambodia access to improved water and sanitation facilities in 1990 were only 35 per cent and 9 per cent as compared to 61 per cent and 29 per cent in 2008. For some countries like Japan, Republic of Korea and Singapore these statistics remain very high at almost 100 per cent throughout the last few decades. This means that in spite of consistent economic growth during the last three decades there are significant disparities among Asian economies in the human development index (HDI).[3] As of 2011, Japan, Republic of Korea and Singapore were ranked 12th, 15th and 26th in the HDI as compared to Nepal, Myanmar, Bangladesh and India which remained in 157th, 149th, 146th and 134th positions respectively (UNDP 2011).

Further, due to growing demand and inefficient use of natural resources, many people in Asia are still living in resource-constrained conditions. In many Asian nations, water, food and energy security is not ensured, and wide income and non-income disparity exists across regions as well as within and between countries. Given the expected growth of economy and population in Asia, resource constraints will become even more acute in the coming years. Recent increases in food and fuel prices have already increased the number of undernourished people from 542 million in 2003 to 583 million in 2007 in the Asia and the Pacific region (FAO 2009a). Between October 2010 and January 2011 alone, the food price index rose by 15 per cent and in early 2011 the domestic prices of rice reached record high levels in some Asian countries such as Bangladesh, China, and Indonesia (FAO 2011). Further, projection shows that the growing food price trend will continue and by 2030 it will increase by some 120 to 180 per cent (Bailey 2011). This will have serious implications for Asia's growth as well as its environment and climate, peace, prosperity and harmony. For this reason there is an urgent need to design and implement strategies that integrate social and environmental concerns into public sector policy making. A resource-constrained future has already sharpened the awareness of policymakers and the general public to the fragility of the existing

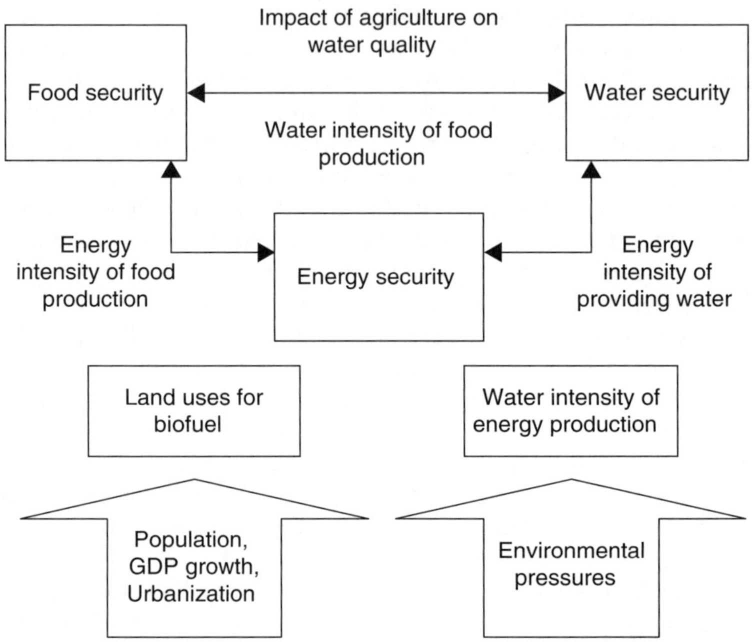

Source: Bentham (2011).

Figure 3.1 Water–energy–food nexus

resources utilization system and has driven all to understand the nexus among interconnected resources and begin adopting strategies to utilize scarce resources more effectively (i.e., adopt effective RCS).

There is also a strong nexus where water, energy and food all combine. For instance, it takes energy to produce, treat and distribute water and it takes water to grow food. Further, agricultural products and water are used for producing energy such as bio-fuel and hydropower. Thus, within the next decades in a resource constraint condition the way we deal with water, energy and food security will significantly influence economic growth, poverty eradication and preservation of important ecosystems (Bonn 2011 Conference 2011). The water, energy and food nexus also has serious implications for the global environment, climate, sustainable growth and human security challenges since a shock to one of these resources will affect production of the others (Figure 3.1).

Figure 3.1 shows how water, food, energy and environment are interconnected. In particular, a substantial volume of water is needed to produce each food item and because of this the agriculture sector uses

70 per cent of the world's withdrawn freshwater resources (World Bank 2008). For example, a single person's balanced diet of 3000 kcal (20 per cent animal and 80 per cent vegetables) requires about 3 to 4 cubic meters of water per day to produce (SIWI 2005). Based on current levels of water productivity, additional use of water for food security by 2025 is estimated to reach 3800 cubic kilometres per year (SIWI 2005). At present, the scarcity of water means that many economies depend on monsoon water for agriculture. This water scarcity is set to worsen with the recent increase of adverse environmental impacts like floods and droughts, which have been the principal reasons for some of the most intensive food emergencies (FAO 2009b). It is also worth noting that uses of water for agricultural activities may also significantly affect the quality of water particularly where proper preventative environmental measures are not taken.

Water also plays a critical role in mitigating energy demands. Globally, the industrial sector uses 16 per cent of total water usage (second largest user of water after agriculture) and the energy sector accounts for 80 per cent of the industrial use (Bonn 2011 Conference 2011). Production of electricity is directly linked to usages of water resources. For example, thermoelectricity generation requires water for mining, processing, cooling and converting primary fuels into electricity. Hydroelectricity generation, processing and refining of liquid fuel production are also highly water consumptive technologies. Besides huge water consumption, these processes of fuel production also cause pollution of water by spills and chemical pollution. On the other hand, water management processes like treatment and distribution need a lot of energy. Further, the energy and food nexus functions both as a demand driver and production input. Both food and energy fulfil basic human needs and also support changes in demand that accompany economic growth. Energy is thus used as an input in food production, processing, and management. On the other hand, agricultural products are used for bio-fuels and other energy production. Between 2007 and 2009, about 10 per cent of global total ethanol was produced from coarse grains, and projections show that by 2030, bio-fuel will power at least 5 per cent of the world's road transportation (IEA 2008). However, this might be an unsustainable trade-off as fields may start being used for fuel production instead of growing food crops. Increased use of food crops for bio-fuel production contributes significantly to food price increases and scarcities in the food and feed markets. This will have a significant impact on many developing countries of Africa and South Asia as a recent study estimates that expansion of bio-fuel production up to 2050 would lead to the number of undernourished pre-school children being 3 million and 1.7 million higher respectively than would have been otherwise the case (FAO 2009b). Thus, within the next decades the way developing Asia deals with water, energy and food security will strongly influence the preservation of important ecosystems, poverty reduction, economic growth, and thus the realization of the Asian century. Therefore, being prepared for a resource scarce future and meeting today's access challenges requires RCS based on regional integration and cooperation that take into account all sides of the water, food and energy nexus.

REGIONAL COOPERATION AND SUSTAINABLE RESOURCE DEVELOPMENT: PRINCIPLES AND DIMENSIONS

Traditionally, in Asia, increasing demand for water, energy and food has been addressed by managing supply-increased production, revamping economic growth and increasing sales and profits of private sector companies. Due to rapid depletion of resources, these traditional practices are not going to be a viable option to manage demand and supply. The individual sectoral approach to water, food and energy security without examining the nexus has inherent limitations as traditional approaches fail to properly recognize the interconnected relationship among these non-renewable public goods and also to take resource efficient strategies for ensuring sustainability. Therefore, it is important to address these issues in tandem and in a comprehensive, systematic manner with appropriate resource centred strategies, policies and planning by proper integration and cooperation at national, sub-regional, regional and global levels.

'Regional integration' here refers to a process of enhancing interconnectivity among economies of a region by enhanced cooperation supported by collective policies and initiatives (ADB 2006). Regional integration is thus broader in scope than regional cooperation and can enhance sustainable management of regional public goods (RPG)/resources by enhancing integrated resource centred strategies. 'RCS' here refers to an efficient and effective way of resource management to ensure a win–win situation for all by maintaining sustainability. Regional integration can enhance RPG management by enhancing RCS supported by proper policies and programmes based on cooperation and comparative advantages.

In a resource constrained world, as countries grow economically, they must implement strategies that include a paradigm of environmental sustainability and optimal use of resources. Enhancing regional cooperation and integration in different layers of resource sharing, production and trade can bind countries into a network of interdependence to ensure maximum use of resources. Thus, liberals, functionalists and neo-functionalists conceive that establishing inter-state linkages in one sector leads to 'spill-over' cooperation in other areas. The functionalist focuses on pursuing cooperation in areas of common interest while leaving areas of opposition and difference untouched initially (Mitrany 1943). According to them, the resultant inter-connectedness would eventually lead to more harmonious inter-state relations. This is in consonance with the liberal argument that trade promotes peace (Gartzke and Li 2003; Mansfield and Pollins 2001; Barbieri 1996; Copeland 1996). However, importantly, while symmetrical trade relations may promote peace, asymmetrical trade that involves dependence of one party on the other may adversely affect the prospects for peace (Balogh et al. cited in Barbieri 1996). Energy trade and investment presents itself as one such area of mutual dependence and common interest in Asia, where inter-state engagement can enhance mutual gain and also lay the ground for interaction in other spheres.

ECONOMIC GROWTH AND ENERGY SECURITY IN ASIA: CHALLENGES AND POTENTIALS

UNDP's (United Nations Development Programme) World Energy Assessment report (UNDP 2004) defines energy security as, 'the continuous availability of energy in varied forms in sufficient quantities at reasonable prices'. Therefore, as Figure 3.2 shows, energy security involves

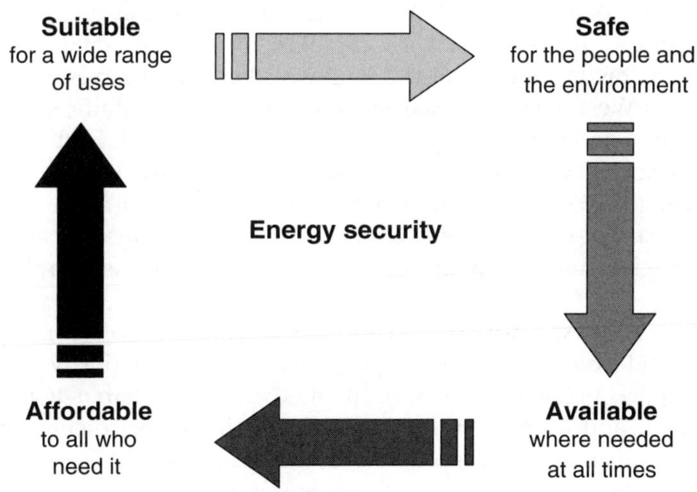

Source: Adapted from ADB (2010).

Figure 3.2 Factors for enhancing energy security

multiple factors, and hence changes in any given factor may help or hurt energy security, depending on the specific circumstances (CSIS 2010). Assessing energy security is complex, and maintenance of energy security for any state has geo-political implications because of the uneven distribution of energy resources across the globe. Factors related to enhancing energy security can be explained by Figure 3.2, which consists of four interrelated factors. In this diagram, the safety and suitability factors are linked to compatibility of energy resources with people and environment, while affordability and availability is linked to reserves as well as patterns of exports and imports of energy resources to meet current and future projected demand.

It is because of this that assessing Asia's energy security also involves considering multiple factors like estimating current energy reserves and analysing patterns of exports and imports, as well as projecting future demand and adopting proper strategies to meet future challenges. Table 3.2 shows that in 2006, developing Asia's total oil reserves were only around 11 000 million tons or around 7 per cent of the world's total. Similarly, gas and coal reserves were around 12 per cent and 32 per cent of the world's total respectively. Region wide analysis shows that, despite it being the most populous region of Asia, South Asia's energy reserves were very low, particularly for oil and gas, as it held 0.5 per cent and 0.8 per cent of the world's total oil and gas reserves. By contrast, East Asia's oil, gas and coal reserve was 1.4 per cent, 1.4 per cent and 13.0 per cent respectively. For Central and West Asia it was 4.1 per cent, 5.6 per cent and 4.6 per cent respectively. Thus, for all regions of Asia, energy reserves were a very low proportion of the total world energy reserves, except for Central and West Asia. This analysis shows unequal distribution of energy in different regions of Asia, as emerging economies of East Asia and populous South Asia's share of energy reserve is much lower compared to Central and West Asia. Therefore, it shows another picture, that Asia has significant scope of regional cooperation and collaboration among its energy deficient and energy sufficient regions to enhance its overall energy security.

Further, East Asia was responsible for the majority of Asia's energy imports, and Southeast Asia was responsible for the majority of Asia's energy exports in 2009 (Table 3.2). In 2009, East Asia shared 46 per cent, 50 per cent, and 65 per cent of Asia's total oil and petroleum, natural gas, and coal imports, respectively. South Asia was the lowest energy exporter in 2009 as it shared only 5 per cent of oil and petroleum and only 1 per cent of Asia's total coal exports. However, during the same period Southeast Asia exported the majority of Asia's oil (51 per cent), gas (54 per cent) and coal (41 per cent) followed by Central and West Asia at 24,

Table 3.2 Asia's proven energy reserves and pattern of energy export and import

Asia's energy reserves total, 2006

Region	Oil		Gas		Coal	
	Million tons	% of world total	Million tons	% of world total	Million tons	% of world total
Developing Asia	11203	7.1	18561	11.6	143051	31.7
East Asia	2219	1.4	2204	1.4	58927	13.0
South Asia	777	0.5	1359	0.8	60843	13.5
Southeast Asia	1665	1.1	5716	3.6	2454	0.5
Central & West Asia	6543	4.1	8890	5.6	20827	4.6
The Pacific	–	–	392	0.2	–	–

Pattern of Asia's energy exports and imports by regions and commodity[a] (%), 2009

Region	Imports			Exports		
	Oil & petroleum products	Natural gas	Coal	Oil & petroleum products	Natural gas	Coal
East Asia	46	50	65	20	1	51
South Asia	18	–	18	5	–	1
Southeast Asia	33	22	14	51	54	41
Central & West Asia	3	28	4	24	45	8

Notes:
– indicates data not available.
Regional aggregates and the world percentage of the world total is calculated based on data for 48 countries that reported.
East Asia includes 6 countries: Hong Kong, China; Japan; Mongolia; People's Republic of China (PRC); Republic of Korea; and Taipei, China.
South Asia includes 8 countries: Afghanistan, Bangladesh, Bhutan, India, Maldives, Nepal, Pakistan, and Sri Lanka.
Southeast Asia includes 10 countries: Brunei Darussalam, Cambodia, Indonesia, Lao People's Democratic Republic, Malaysia, Myanmar, Philippines, Singapore, Thailand, and Vietnam.
Central West Asia includes 8 countries: Armenia, Azerbaijan, Georgia, Kazakhstan, Kyrgyz Republic, Tajikistan, Turkmenistan, and Uzbekistan.
The Pacific includes 14 countries: Cook Islands, Democratic Republic of Timor-Leste, Federated States of Micronesia, Fiji Islands, Kiribati, Marshall Islands, Nauru, Papua New Guinea, Republic of Palau, Samoa, Solomon Islands, Tonga, Tuvalu, and Vanuatu.
[a] As a percentage of total Asian exports and imports of that commodity.

Source: World Resources Institute (2009).

Table 3.3 Asian energy demand and supply projection

	1990	2000	2007	2030	2050
Asia energy demand (Mtoe)	2220	2910	4242	7980	11480
China	872	1105	1970	3637	5011
India	318	457	622	1341	2389
ASEAN	243	389	513	903	1177
Central Asia	198	128	159	256	385
High income Asia*	629	746	896	995	1112
Asia electricity consumption (TWh)	2249	3057	6113	17267	26181
China	586	1081	2717	7513	10630
India	197	369	544	1966	3440
ASEAN	167	321	497	1383	1956
Central Asia	162	124	152	443	715
High income Asia	976	1012	1128	1411	1746
Asia energy supply mix (%)					
Coal	40	42	47	48	50
Oil	16	17	20	21	20
Gas	9	10	11	12	11
Hydro	3	2	2	2	1
Biomass	26	24	15	10	7
Other (including nuclear)	6	5	5	7	11

Notes:
* As defined by the World Bank, High Income Countries are economies of $12276 or more according to 2010 gross national income (GNI) per capita.
Mtoe = Million tons of oil equivalent.
TWh = Tera-watt-hour.

Sources: EIA (2010); IEA (2008, 2009); IEA (2010a, 2010b); World Bank (2010).

45 and 8 per cent respectively. This analysis also shows that Asian countries have ample scope to enhance regional trade. This increased regional energy trade can enhance Asia's overall energy security by ensuring safe, available, affordable and suitable sources of energy from the regional level.

Further analysis for Asia's future energy demand projections show the emerging economies of Asia, led by China and India, will drive world energy demand, due to increasing demand for both fossil fuels and renewable sources of energy (Table 3.3). Asian total energy consumption increased from 20 per cent in 2000, to 27 per cent of global total energy consumption in 2007, and is expected to reach 40 per cent by 2050. By 2030, developing Asia's energy needs will have grown at more than twice the world average, accounting for the additional global supply of almost all coal, three-fifths of oil, and a third of natural gas destined for developing countries – and almost a third of world energy consumption (ADB 2007). Another estimates shows by 2030, developing Asia will be consuming about 26 per cent of the global demand and the region's dependence on imported oil will increase from 43 per cent in 2002 to 78 per cent in 2030 (IEA and OECD 2004).

Table 3.3 shows that Asia's demand for both fossil fuel and electricity will be driven mainly by the increasing demand of China and India. China already surpassed the US in 2010 to become the largest energy consuming country in the world (IEA 2010a). Moreover, before 2030 Asia as a whole will surpass the OECD to become the largest energy-consuming block. China's total energy demand is expected to increase around five times to reach around 5011 Mtoe in 2050 from around 870 Mtoe in 1990. Similarly, India's total energy demand will also increase around seven times their current demand during the same period. However, demand for consumption of electricity in Asia will increase much faster, around 19 times current demand, and is estimated to reach around 24000 TWh in 2050 from around 1200 TWh in 1990, primarily led by the demand from China and India (IEA 2010b).

On the other hand, the projections for Asia's energy supply mix show that the contribution from biomass will reduce significantly from 26 per cent in 1990 to 7 per cent in 2050. However, dependency on nuclear, gas, oil and coal is expected to increase further (Table 3.3). Asia's overall oil import dependency also shows significant increase in the decades ahead as it is expected to go from 44 per cent in 2008 to 75 per cent in 2030 and 90 per cent in 2050 (EIA 2010).

The above analysis of Asia's energy reserves, export and import patterns, and future demand projections reveal important aspects for enhancing Asia's energy security to sustain its rapid economic growth. First, Asia's energy reserves are not only insufficient but also unevenly distributed among regions. Asia's future growth drivers, China and India, show insufficient energy reserves to sustain projected economic growth, whereas Central and West Asian developing countries have higher energy reserves. Second, Asia's future energy demand will increase significantly, particularly led by the economic growth of China, India, ASEAN and South Asia. Third, the energy supply mix shows Asia's increasing dependence on fossil fuel and decreasing share of biomass to meet future energy demand. Therefore, increasing regional cooperation and adopting integrated resource centred strategies at national, sub-regional and regional levels can enhance Asia's energy security to realize the Asian century by achieving sustained and sustainable economic growth.

REGIONAL COOPERATION AND ENERGY SECURITY IN ASIA: SCOPE AND OPPORTUNITIES

Asia has energy surplus and energy deficit countries. Increasing regional cooperation in the energy sector and adopting resource-centred strategies for the management of related natural resources can enhance energy security for the whole region so everybody can win. For example, development of greener cross-border energy projects can allow countries to benefit from natural endowments, providing efficient and secure supplies of electricity, coal, gas, oil, and alternative energies (ADB/ADBI 2009).

For energy surplus countries like Bhutan, Cambodia, Kyrgyz Republic, Lao PDR, Myanmar, Nepal, Tajikistan, and Turkmenistan, regional energy trade can result in huge economic gains. In the 2007 fiscal year Bhutan's electricity export contributed nearly 25 per cent of GDP and 60 per cent of government revenues. On the other hand, for energy deficient countries like Afghanistan, Bangladesh, China, India, Pakistan, and Sri Lanka, importing energy will allow their economies to grow faster. Further, cross-border energy projects involving hydro, nuclear, and wind power could also reduce electricity generation from coal and oil and thus limit environmental damage. This is particularly important for countries like India and China which emit a lot of CO_2. China's CO_2 emissions increased rapidly since 2001, while the emissions intensity decreased (21 per cent in 2008). In India, CO_2 emissions continue to grow since 2003 (5 per cent in 2008) and among ASEAN countries Indonesia, Thailand and Malaysia are the top three CO_2 emitters.

Thus, by regional cooperation, both energy surplus and energy deficient countries can enhance national energy security by diversifying energy forms and supply sources and lowering costs. Studies estimate that by enhancing regional cooperation, the Greater Mekong Sub-region (GMS) can reduce the sub-region's total discounted energy costs by an estimated $220 billion, or 19 per cent of total energy costs (ADB 2011). Likewise many Asian countries have unused savings and reserves, which can be channelled to finance energy and other infrastructure projects at both the national and the regional levels. Thus, the development of stable and efficient regional financial markets can channel savings from around Asia and the rest of the world into productive investments, notably infrastructure, throughout the region (ADB/ADBI 2009).

For energy security in Asia another important issue is reducing the dependence on fossil fuel. Globally China (3163Mt) and India (538Mt) are considered as the largest and third largest coal producers respectively. In 2008 coal was the major source of energy consumption in China (71 per cent), and India (42 per cent); oil was the main source in Indonesia

(35 per cent), Thailand (41 per cent) and Singapore (62 per cent); and gas for Malaysia (51 per cent) and Brunei (79 per cent) (IEA 2011). China's industry is the largest energy-consuming sector, accounting for 46 per cent of total final energy consumption, and for India it is the residential sector, accounting for 34 per cent of total final energy consumption. These countries need to shift their dependence from fossil fuels to non-fossil fuels by diversifying their energy production and enhancing use of renewable energy. Due to increased scarcity of fossil fuels, hydropower can be the largest source of renewable and new energy for many Asian countries. China and India have relatively fast development of wind and nuclear energy; Philippines and Indonesia are rich in geothermal resources and Thailand's biomass already accounted for a large proportion of its new energy market. Similarly, Japan also has constraints for fossil fuel, but has huge potential for renewable energy such as wind and hydropower, particularly on small islands.

Considering the abundant availability of various sources of renewable energy such as solar, wind, biomass and ocean, cooperation in the field of renewable energy has huge potential in Asia. Among South Asian countries India has already developed its manufacturing capability in wind power generators as well as development of biomass gasifiers and cook stoves (Nanda and Goswami 2008). On the other hand countries like Pakistan, Sri Lanka and Bangladesh have good wind power potential. Therefore, cooperation for technology transfer from India to these countries can enhance South Asia's energy security by using renewable energy.

Similarly, the ASEAN region, particularly the Greater Mekong Subregion (GMS), which includes Cambodia, Lao, Myanmar, Thailand and Vietnam, as well as Malaysia and Indonesia, has huge potential for cooperation for hydropower development, and Indonesia and the Philippines have abundant geothermal energy. Moreover, among Asian countries China has already become a global leader in manufacturing and development of various renewable energy related technologies such as wind turbines, solar cells etc. Therefore, technology transfer from China to South Asia and ASEAN countries can significantly increase production of renewable energy in these regions.

Asia's energy demand, supply and energy mix scenario shows energy self-sufficiency in most countries will decline in 2030 and their dependence on imported energy will increase. China's energy self-sufficiency[4] will decline to 68.5 per cent, as it becomes a net coal importer and its dependence on imported oil and gas will rise to 70 per cent and 71 per cent, respectively. India's dependence on imported energy will rise and energy self-sufficiency will decline to less than 50 per cent. Among ASEAN countries, only Brunei remains fully energy self-sufficient. Indonesia, Malaysia and Vietnam will turn into net oil importers in 2030 (IEA 2010c). For this reason, to secure energy security Asian countries need to enhance regional cooperation and collaboration. Asian countries could learn from their national indigenous knowledge, regional best practices and global efficient examples to enhance efficiency, effectiveness and cooperation in energy generation, transition and power trade.

From the national level in Asia, one such indigenous best practice example is 'A Liter of Light Project' in the Philippines.[5] In this simple resource efficient energy project, used plastic bottles are up-cycled into solar bottle bulbs by My Shelter Foundation and various municipalities. Thus, in total 15 000 solar bottle bulbs have been installed in 20 cities. In every bottle, saving in electricity expenditure is at an average of US$6.00/month. Thus this project provides free energy and there are no carbon emissions. Moreover, this is immediately scalable by social enterprise and easily replicated by the local government.

Another such best practice example of energy efficient public–private partnership is the 'Solar LED Lighting 1000 Villages Program' in China.[6] In this project the Climate Group, the One Foundation and Philips lighting have partnered with the city of Guiyang to collaborate on the installation of LED street lighting. This provides a high-quality, sustainable, cost-efficient lighting using sunlight for people in remote areas who do not have access to the conventional electricity grid. Its carbon emissions during use are zero. Thus, over 1.3 million tons of carbons are reduced. Under this project there are 400 demo villages in China in the first two years starting from August 2009 and another 600 villages in China, India and African countries in the following three years.

From global best practices, one resource efficient strategy that can be replicated in many countries of developing Asia is introducing clean cook stoves. In Africa, under the Global Alliance for Clean Cook stoves, private sector partners are working with multiple national governments and the UN Foundation to distribute efficient cook stoves named 'Rocket Stoves'.[7] These clean stoves are 40–60 per cent more efficient than fire stoves and per stove reduce 1–2 metric tons of carbon per year. Clean stoves also reduce black carbon and indoor air pollution.

At the regional level within Asia, ASEAN has good examples of internal energy cooperation such as the ASEAN Petroleum Security Agreement, Trans-ASEAN Gas Pipeline Project, and ASEAN Power Grid Project. However, currently, regional cooperation in the energy sector in Asia has concentrated in a market-based resource allocation and optimization and enterprise energy exploration and development, while the building for inter-governmental dialogue and cooperation mechanism of energy is still relatively weak. Considering existing and potential multidimensional challenges in the future, cooperation should be strengthened in, for example, new and renewable energy, energy efficiency, environmental protection, scientific research and technology development.

Challenges for Energy and Related Infrastructure Sector Development in Asia

Overall analysis shows that for a resource constrained future, isolated policy action at the national or sub-national/regional levels cannot resolve many issues related to Asia's energy security. Therefore, to sustain its economic growth for enhancing energy security, Asia needs cross-border coordination between national governments, collaboration between national and local governments, as well as creation of institutions that can support and enable consultation with stakeholders and include their views/concerns. Asia needs to forge strategic alliances at regional levels to develop consolidated approaches and overcome various challenges related to the energy sector, such as lack of private participation, limited resources in the low income countries (LICs), lack of regional cooperation, and policy and institutional weakness.

Lack of private participation

Infrastructure projects are usually expensive and lengthy, hence recouping financial returns also takes a long time. They also involve risks associated with uncertainties about future costs and revenues. Consequently, private parties are usually reluctant to invest in infrastructure. As Figure 3.3 shows, during the period 1990–2008 China received the highest private sector involvement in terms of both number of projects and total amount invested, followed by India. This amount is much lower in comparison to many other countries in the same stage of their economic development. Therefore, to encourage private investors to fund and provide efficient infrastructure facilities Asian countries need to create efficient and effective harmonious policies and procedures at the national, sub-regional and regional levels (ADB/ADBI 2009).

Limited resources and poor governance in the low income countries

Asian low income countries[8] have limited fiscal space to address the growing needs of infrastructure financing as well as maintenance. Investment needs as percentage of GDP show that it can be a real burden for the poorer Asian countries. Further, in many developing countries maintenance costs for existing infrastructure can consume as much as 70 per cent of total investment needs, as is the case for Bhutan. In Nepal during 1990–2007, government spending on electricity, gas, and water as a

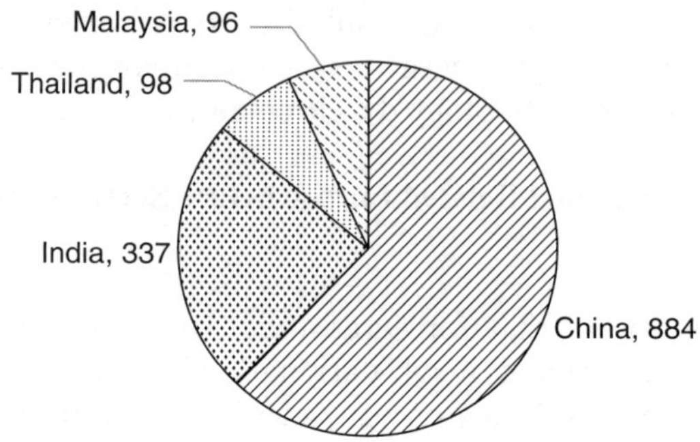

(a) Top Asian countries by number of projects: 1990–2008

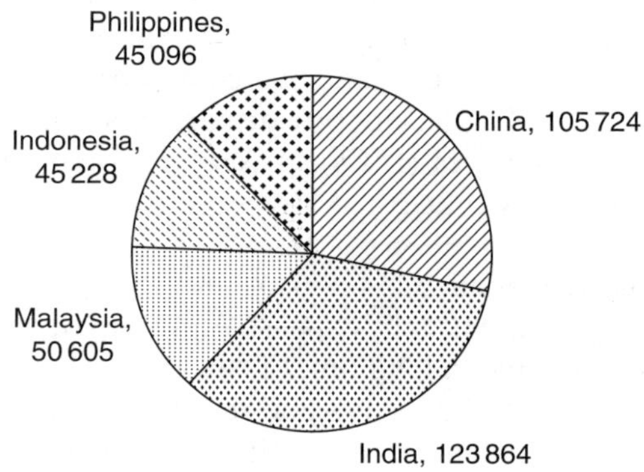

(b) Top Asian countries by value: 1990–2009 (US$ million)

Sources: World Bank and PPIAF (various years).

Figure 3.3 Private sector infrastructure investments in Asia, 1990–2008

percentage of total government expenditures declined by 58 per cent, and for transport by 32 per cent. To increase its electricity generation capacity Nepal would require an investment of about US$1.5 billion in 2007 prices, and to raise both the electrification rates and availability of power to the consumption levels that prevail in India would cost an estimated US$5.1 billion in 2007 prices (ADB, DFID and ILO 2009). To meet demand, governments of Asian developing countries need to commit a greater portion of their national GDP to financing energy sector investments than current levels.

Further, political risk and governance issues can not only limit the possibility of attracting private funding in infrastructure but also reduce the value of infrastructure by distorting design and construction as well as by raising the cost (Tanzi and Davoodi 1998). Similarly, due to poor governance and accountability mechanisms, developing Asia has not only failed to attract enough private funding in the energy sector but also still lags behind in terms of infrastructure quality and competitiveness. In 2011 in overall quality of infrastructure among 142 countries of the world, the two Asian giants India and China ranked 86th and 69th respectively. Nepal was in 132nd position, Bangladesh 129th and Vietnam 123rd (WEF 2011). Therefore, besides increasing financial resources, public policies in developing Asia should focus on improving governance in the infrastructure sector in general and in the energy sector in particular by enhancing accountability and transparency.

Lack of regional cooperation, and policy and institutional weaknesses

The effective management of energy resources is inter-disciplinary by nature and it requires regional cooperation and coordination of policies and institutions. However, developing Asia lacks significantly in cooperation and policy and institutional effectiveness. According to a World Bank report published in 2007, South Asia is the least integrated region in the world[9] (Ahmed and Ghani 2007). The institutional framework in the South Asian countries is both inadequate and fractured. Further, in different parts of Asia there are many overlapping sub-regional institutions operating with varying speeds and addressing regional infrastructure issues in different degrees, with multiple objectives (Figure 3.4). Most sub-regional institutions are informal (except ASEAN and South Asian Association for Regional Cooperation (SAARC)), with no legal binding or enforcement capacity, and most take advisory, regulatory, and financing modalities. Therefore, from both the national and regional institutional levels Asian countries face multiple challenges for infrastructure and energy sector development, such as: insufficient market opening and capital account liberalization; limited and varying degrees of transparency, financial regulation, financial supervision, and corporate governance; inadequacies in risk management in financial firms and markets; and heterogeneity of supervisory, accounting and auditing rules and regulatory frameworks across countries (ADB 2007).

Investment Needs for Asia's Infrastructure and Energy Sector Development

Between 2010 and 2020, Asia needs to invest about US$8.0 trillion (about US$750 billion per year) in both national and regional infrastructure for energy, telecommunications, transport, water and sanitation (ADBI 2009)

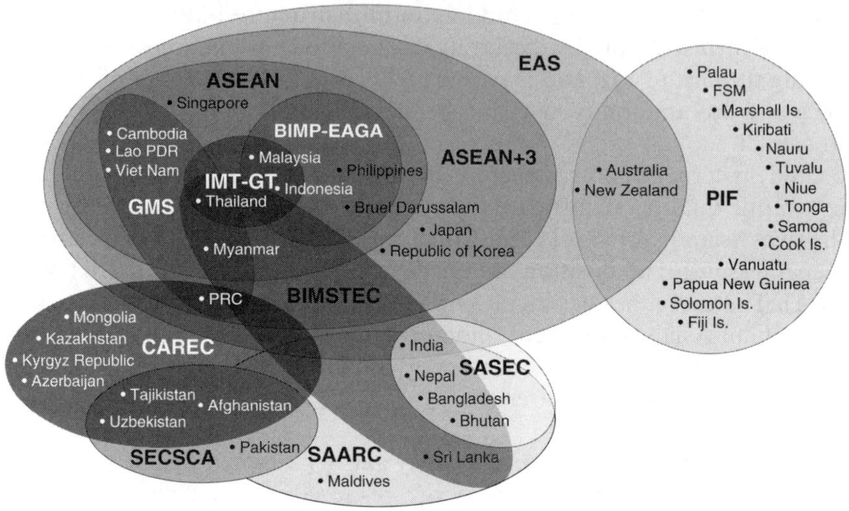

Notes:
ASEAN = Association of Southeast Asian Nations.
ASEAN+3 = ASEAN and Brunei Darussalam, Japan, Republic of Korea.
BIMP–EAGA = Brunei Darussalam, Indonesia, Malaysia, Philippines – East ASEAN Growth Area.
BIMSTEC = Bay of Bengal Initiative for Multi-Sectoral Technical and Economic Cooperation.
CAREC = Central Asia Regional Economic Cooperation.
EAS = East Asia Summit.
FSM = Federated States of Micronesia.
PIF = Pacific Island Forum.
SAARC = South Asian Association for Regional Cooperation.
SASEC = South Asia Sub-regional Economic Cooperation.
SECSCA = Sub-regional Economic Cooperation in South and Central Asia.

Source: ADB/ADBI (2009).

Figure 3.4 Architecture of sub-regional infrastructure cooperation in Asia

(Table 3.4). According to rough estimates, some US$17 trillion in energy sector investments is required globally by 2030, half of it in Asian developing countries, to expand energy production and distribution facilities and infrastructure in order to meet additional baseline demand across all fuel categories – which is projected to be much higher than the corresponding increase witnessed over the previous three decades (ADB 2007). However, regional resources are invested elsewhere, and infrastructure projects in developed countries attract far greater global private sector funding than developing economies in the region. The private sector share in East Asia's infrastructure investment is quite low – about 20 per cent prior to the 1997

Table 3.4 Financing needs for Asia's national infrastructure (2008 US$ million), 2010–20

Sector/Subsector	New capacity	Replacement	Total
Energy (electricity)	3176437	912202	4088639
Transport	1761666	704457	2466123
Telecommunications	325353	730304	1055657
Water and sanitation	155493	225797	381290
Total	5418949	2572760	7991709

Source: ADBI (2009); Centennial Group Holdings (2009).

Asian economic crisis. Therefore, to meet huge infrastructure investment needs in decades ahead, Asia's developing countries have to open their economies to greater foreign direct investments (FDI), ensure economic and political stability, and provide incentives to offset perceived political, security, and financial risks (ADB 2007).

Potential Sources of Financing in Energy Sector in Asia

Financing huge energy infrastructure needs at national and regional levels is a considerable challenge, particularly for those developing Asian countries with inadequate fiscal space. While the public sector will shoulder the bulk of financing for new energy projects and repair/maintenance projects, an effective private sector will be necessary to fill remaining financing gaps. Asian countries therefore need to create policies and procedures that encourage private investors to fund and provide efficient infrastructure facilities (ADB/ADBI 2009). Thus, Asian energy sector financing can be supported by four different sources:

1. Public sector

As in most developed and developing countries of the world, traditional public sector strategies such as policies, funding, and fiscal stimulus are still the major source of financing as well as increasing efficiency and effectiveness in the energy sector in many developing countries of Asia. For example in developing countries like Nepal and Afghanistan, over 90 per cent of domestic infrastructure investment is financed from domestic financial markets or through government budgets funded by local taxes (ADB/ADBI 2009). Similarly, in India about 75 per cent and in China most energy sector institutions are state owned. To enhance energy security under the 12th Five Year Plan (2011–15) the Chinese government took

Table 3.5 Financial resources in Asia: savings and reserves

Country/region	GDP	Gross savings 2009		Foreign exchange reserves 2009	
	US$ billion	US$ billion	% of GDP	US$ billion	% of GDP
PRC	4985	2355*	47	2399	48
Hong Kong, China	n.a.	64*	–	256	–
Japan	5068	n.a.	–	997	20
Korea, Republic of	833	286*	34	265	32
Indonesia	540	83	15	61	11
Malaysia	192	n.a.	–	93	48
Philippines	160	26	16	38	24
Singapore	182	87*	48	186	102
Thailand	264	80*	30	134	51
India	1310	410*	31	259	20

Notes:
n.a. = not available.
* 2008 data.

Sources: World Bank (2010); IMF–IFS (2010).

specific measures such as reduction in energy intensity by 16 per cent, and increasing the share of renewables in primary energy consumption to 11.4 per cent from 2011 levels of 8.3 per cent (State Council of China 2011). Similarly, to enhance energy security, in addition to various financing mechanisms, policies by the Indian government gave emphasis to controlling losses in the power transmission and distribution networks (Planning Commission of India 2006). Further, in developing Asia, despite constraints public financing comes from sources such as savings, sovereign wealth funds and foreign reserves (Table 3.5). The share of national financing of energy investments needs to be increased from the historical average of 1.0–1.5 per cent of GDP witnessed for most developing Asian countries (ADB 2007). As data in Table 3.5 show, many Asian countries have considerable gross savings and foreign exchange reserves. Therefore, Asian countries can take cooperative measures to use Asia's savings and reserves to meet Asia's energy infrastructure financing demand.

In addition, by enhancing transparency and accountability by proper policies, the public sector can provide significant opportunities to increase both public and private investments in the energy sector. However, public financing in the energy sector should not focus just on installing decentralized technologies but should also take proper policies to enhance social investment by undertaking considerable planning, consultation and work. Public policies should be targeted to explore innovative financing instruments and accelerate R&D support for enhancing efficiency and effectiveness of both the supply and demand sides of the energy sector.

2. Private sector and PPPs

In the energy sector different players have unique roles to play, as they have different risk and required return preferences. In developing Asia, the energy sector needs increasing private sector participation for the development of technology, skills and management capability as well as the transfer of technological know-how that the public sector does not possess. To attract private institutions simplicity, clarity, and consistency around policy and support mechanisms are key. Private sector and PPPs can be engaged by mobilizing private portfolio funds, increasing FDI, joint ventures, concessions, management contracts, BOO (Build Own Operate), BOT (Build Operate and Transfer) and BOOT/BOLT (Build Own Operate Transfer/Lease Transfer) (DFAT 1998). The public sector needs to create an environment suitable for participation of the private sector and PPPs in national and regional energy projects.

3. Regional savings and financing mechanisms

Asian financial market integration could enhance cross-border capital and utilize Asia's robust savings/foreign exchange reserves for national and regional infrastructure investment. Several Asian countries, such as China, India, Japan and Republic of Korea, have huge accumulated foreign exchange reserves which are mostly invested in 'risk-free' US and European government securities (ADB/ADBI 2009). Taking proper measures to utilize Asia's savings for its own infrastructure and energy sector development can enhance the financial capability of poor countries. At the regional level enhancing mechanisms like micro financing, especially for renewable energy-based supplies, concessional loans, and grants can be effectively packaged with innovative marketing schemes (ADB/ADBI 2009). Moreover, regional financing for the energy sector can be promoted by setting up a regional infrastructure development fund or issuing regional infrastructure or energy bonds (ADB 2008b).

4. Multilateral and bilateral institutions

For promoting regional cooperation and enhancing energy security in Asia, multilateral and bilateral institutions such as the Asian Development Bank (ADB), Japan International Cooperation Agency (JICA) and World Bank (WB) can play multidimensional roles. For example they can expand lending programmes to assist in rapid delivery of energy projects; offer an innovative range of credit enhancements and risk mitigation measures; help create bankable projects, mobilize long-term funds through capital markets, and arrange co-financing (Kuroda et al. 2008). They can also help to create appropriate and innovative financial instruments for PPP projects to encourage private sector investment; promote further integration and enhancement of Asian financial market efficiency, liquidity, and depth; act as knowledge partners, technical advisers, capacity builders, and honest brokers; strengthen policy guidance and bring additional financial and technical expertise; and promote adherence to international and regional standard best practices.

As a major institution in Asia, the ADB is helping to improve the flow of private savings and capital into productive investments, including infrastructure, by developing bankable projects, designing innovative financial instruments, enhancing financial markets, and promoting further financial integration. By its 'Strategy 2020', the ADB is promoting development of its member countries by three pillars: inclusive economic growth, environmentally sustainable growth, and regional integration (ADB 2008a). It has four core areas of operation which either directly or indirectly facilitate development and financing of energy and other infrastructure sectors.

- The infrastructure core focuses both on building physical assets and related institutional/policy reforms, and aims at promoting investments in improving transport, communication, and energy connectivity within and between developing member countries.
- The environment core seeks to ensure sustainable development.
- The regional cooperation and integration core focuses on increasing financial resources for regional programmes and projects including regional infrastructure development and connectivity, and supporting developing member countries and regional bodies to build their institutional capacity to manage regional cooperation and integration.
- The financial development core supports the development of financial infrastructure, institutions, and products and services; facilitates the creation of a harmonized regulatory framework, a common credit guarantee mechanism, and a regional clearing and settlement system; and promotes channelling the region's savings into the most productive investments, particularly development of the capital market.

CONCLUSION AND FUTURE RESEARCH DIRECTIONS

From the overall analysis it is evident that developing Asia lacks conceptual frameworks as well as practical action frameworks at the national, sub-regional, and regional levels for enhancing long term energy security. Considering resource constraints and great diversity among developing Asian countries for enhancing energy security this study proposes a mixture of frameworks by the active participation of both private and public sectors at different levels (see Ramani (2009)). Such a framework is mainly based on the principle of enhancing regional cooperation for sharing both 'soft (knowledge, R&D, technology etc.) and hard (money and other physical resources) infrastructure' in the energy sector. However, it also recognizes the importance of increasing capacity at the national and sub-regional levels by improving governance and sharing knowledge and technology. Thus, considering a resource constraint future scenario it gives more flexibility to promote resource centred strategies for enhancing energy security in Asia.

However, by taking evolving opportunities for materializing the proposed energy sector cooperation framework, Asia needs to address multiple related challenges. Among multiple priorities for enhancing long term energy security Asia needs to evolve a suitable and robust regional financing mechanism. Though the public sector will continue to play a major role in funding energy infrastructure investments, private sector participation will be increasingly important. Strengthening of existing and development of new national and regional institutions for investment in innovative energy infrastructure development would increase policy effectiveness, create more effective markets and – more importantly – create a financial system that could be stronger, more integrated and more competitive. Further, developing innovative and effective financing instruments with appropriate incentives, particularly for the private sector, is important not only to increase the use of regional resources for Asian investment priorities but also to attract investments from other parts of the world such as Europe and the US. Lastly, support from multilateral and bilateral development institutions must be increased by adopting new and innovative assistance strategies with greater emphasis on infrastructure investments.

NOTES

1. For the purpose of this study, developing Asia is defined as all 43 developing member countries of the Asian Development Bank (ADB), see http://www2.adb.org/documents/reports/global-food-price-inflation/food-price-inflation.pdf.

2. Emerging Asia includes People's Republic of China (PRC), Japan, Republic of Korea and major ASEAN economies (Indonesia, Malaysia, Philippines, Singapore, Thailand, and Vietnam), see: http://aric.adb.org/pdf/workingpaper/WP31_Decoupling_or_Recoupling.pdf.
3. The HDI is a composite index giving equal weights on three dimensions and four indicators of human well being, namely living standards (GNI per capita), health (life expectancy at birth), and education (mean years of schooling and expected years of schooling).
4. Energy self-sufficiency is expressed as a ratio between domestic production and total consumption.
5. For details see http://isanglitrongliwanag.org/.
6. For details see http://community.lighting.philips.com/blogs/asimpleswitch.
7. For details, see Global Alliance for Clean Cookstoves, http://cleancookstoves.org/.
8. LICs of $1005 or less according to 2010 GNI per capita as divided by the World Bank.
9. Integration is measured by 'intra-regional trade in goods, capital and ideas' (Ahmed and Ghani 2007).

REFERENCES

ADB (Asian Development Bank) (2006), *Regional Cooperation and Integration Strategy*, Manila: ADB.

ADB (2007), *Energy for All: Addressing the Energy, Environment, and Poverty Nexus in Asia*, Manila: ADB.

ADB (2008a), *Strategy 2020: The Long-term Strategic Framework of the Asian Development Bank 2008–2020*, Manila: ADB.

ADB (2008b), 'Asian infrastructure financing initiative', Board Information Paper, Manila: ADB, August.

ADB (2009), *Asian Development Outlook 2009: Rebalancing Asia's Growth*, Manila: ADB.

ADB (2010), 'Development Asia', July–September, Manila: ADB, http://www.development.asia/issue11/default.asp, accessed 20 March 2012.

ADB (2011), 'Key indicator for Asia and the Pacific 2011', Manila: ADB, http://www.adb.org/Documents/Books/Key_Indicators/2011/pdf/Key-Indicators-2011.pdf, accessed 21 March 2012.

ADB/ADBI (Asian Development Bank Institute) (2009), *Infrastructure for a Seamless Asia*, Manila: ADB and Tokyo: ADBI.

ADB, DFID (Department for International Development) and ILO (International Labor Organization) (2009), *Nepal: Critical Development Constraints, Country Diagnostics Studies Highlights*, Manila: ADB; London: DFID; Geneva: ILO.

ADBI (Asian Development Bank Institute) (2009), 'Demand for infrastructure financing in Asia 2010–2020', *ADBI Internal Report* (prepared by Centennial Group Holdings, LLC, Washington, DC), Tokyo: ADBI.

Ahmed S. and E. Ghani (2007), 'South Asia's growth and regional integration: an overview', in S. Ahmed and E. Ghani (eds), *South Asia: Growth and Regional Integration*, Delhi: World Bank/Macmillan, pp 3–42.

Bailey, R. (2011), 'Growing for a better future: food justice in a resource-constrained world', London: Oxfam, http://www.oxfam.org/sites/www.oxfam.org/files/growing-a-better-future-010611-en.pdf, accessed 15 February 2012.

Barbieri, K. (1996), 'Economic interdependence: a path to peace or source of interstate conflict?', *Journal of Peace Research*, **33** (1), 29–49.

Bentham, J. (2011), *Water, Energy, and Food Security in the Urban Context*, presentation from the 2011 World Water Week in Stockholm, http://www-static.shell.com/static/aboutshell/downloads/our_strategy/shell_global_scenarios/speech_jeremy_bentham_world_water_water_24082011.pdf, accessed 10 December 2011.

Bonn 2011 Conference (2011), 'The water, energy and food security nexus: solutions for the green economy', 16–18 November, http://www.water-energy-food.org/, accessed 10 February 2012.

Centennial Group Holdings (2011), Centennial Group Projections 2011 in H. Kohli, A. Szyf and D. Arnold (2011), 'Possibilities for the world's economies through 2050 using a revised growth model for 185 countries', *Global Journal of Emerging Market Economies, Emerging Markets Forum*, January, **3** (1), 59–110.

Centennial Group Holdings (2009), 'Estimating infrastructure demand for Asia and the Pacific 2010–2020', *ADBI Commissioned Report*, Tokyo: ADBI.

CSIS (Center for Strategic and International Studies) (2010), 'Defining energy security factors', Washington, DC: CSIS, http://csis.org/files/media/csis/pubs/090130_Annex_II_Defining_Energy_Security_Fac.pdf.

Copeland, D.C. (1996), 'Economic interdependence and war: a theory of trade expectations', *International Security* **20** (4), 5–41.

DFAT (Department of Foreign Affairs and Trade) (1998), 'Asia's infrastructure in the crisis: harnessing private enterprise', Australian Government, http://www.dfat.gov.au/publications/catalogue/Infrastructure.pdf.

EIA (Energy Information Agency) (2010), *China Energy Data, Statistics and Analysis – Oil, Gas, Electricity, Coal*, Washington, DC: EIA.

FAO (Food and Agriculture Organization) (2011), 'Global food price monitor', Rome: FAO, http://www.fao.org/giews/english/gfpm/GFPM_02_2011.pdf, accessed 19 February 2012.

FAO (2009a), 'Hunger on the rise – soaring food prices and 75 million people to global hunger rolls', Briefing Paper, 17 September 2008, cited in *Economic and Social Commission for Asia and the Pacific, Economic and Social Survey of Asia and Pacific 2009: addressing triple threats to development* (ST/ESCAP/2522), Bangkok: FAO, http://www.unescap.org/survey2009/download/Survey2009.pdf, accessed 19 February 2012.

FAO (2009b), 'How to feed the world in 2050', Rome: FAO, http://www.fao.org/fileadmin/templates/wsfs/docs/expert_paper/How_to_Feed_the_World_in_2050.pdf.

Gartzke, E. and Q. Li (2003), 'War, peace and the invisible hand: positive political externalities of globalization', *International Studies Quarterly*, **47** (4), 561–86.

IEA (International Energy Agency) (2011), *World Energy Outlook 2011: Are We Entering a Golden Age of Gas?*, Paris: IEA.

IAE (2008), 'Global energy trends to 2030', Paris: IEA, http://www.iea.org/weo/docs/weo2008/WEO_2008_Chapter_8.pdf.

IAE (2009), 'World energy outlook, 2009', Paris: IEA, http://www.worldenergyoutlook.org/.

IAE (2010a), 'World energy outlook 2010', Paris: IEA, http://www.worldenergyoutlook.org/.

IAE (2010b), 'China overtakes the United States to become world's largest energy consumer', Paris: IEA, http://www.iea.org/index_info.asp?id=1479.
IAE (2010c), 'Energy technology perspectives', Paris: IEA, http://www.iea.org/techno/etp/index.asp.
IAE and OECD (Organisation for Economic Co-operation and Development) (2004), *World Energy Outlook 2004*, Paris: IEA.
IMF (International Monetary Fund) and IFS (International Financial Statistics) (2010), 'IMF–IFS database', Washington, DC: IMF, http://www.imf.org/external/data.htm, accessed 9 September 2011.
Kohli, Harinder S., A. Sharma and A. Sood (eds) (2011), *Asia 2050: Realizing the Asian Century*, Sage Publications Pvt Ltd, Manila: ADB.
Kuroda, H., Kawai, M. and R. Nangia (2008), 'Infrastructure and regional cooperation', in F. Bourguignon and B. Pleskovic (eds), *Rethinking Infrastructure for Development*, Washington, DC: World Bank, pp. 235–60.
Mansfield, E.D. and B.M. Pollins (2001), 'The study of interdependence and conflict: recent advances, open questions, and directions for future research', *Journal of Conflict Resolution*, **45** (6), 834–59.
Mitrany, D. (1943), *A Working Peace System: An Arrangement for the Functional Development of International Organization*, London: Royal Institute of International Affairs.
Nanda, N. and A. Goswami (2008), 'Energy cooperation in South Asia: prospects and challenges', *Man and Development*, **XXX** (1).
Planning Commission of India (2006), *Towards Faster and More Inclusive Growth: An Approach to the 11th Five Year Plan 2007–2012*, New Delhi: Oxford University Press.
Ramani, K.V. (2009), 'Asia-Pacific energy security cooperation framework initiative', www.un.org/regionalcommissions/2ndcom09present/escap.ppt.
SIWI (Stockholm International Water Institute) (2005), 'Let it reign: the new water paradigm for global food security', Stockholm: SIWI, http://www.waterfootprint.org/Reports/SIWI_2005.pdf, accessed 15 February 2012.
State Council of China (2011), 'The 12th five year plan of national economy and social development of People's Republic of China 2011–2015', Beijing.
Tanzi, V. and H. Davoodi (1998), 'Corruption, public investment and growth', in H. Shibatu and T. Ihori (eds), *The Welfare State, Public Investment and Growth*, Tokyo: Springer-Verlag, pp. 41–60.
UN (United Nations) (2011), *World Population Prospects: The 2010 revision*, New York: UN.
UNDP (United Nations Development Programme) (2011), 'Human Development Report 2011: sustainability and equity: a better future for all', New York: UN, http://hdr.undp.org/en/statistics/hdi/, accessed 19 February 2012.
UNDP (2004), 'World energy assessment', New York: UN, http://www.undp.org/energy/docs/WEAOU_full.pdf, accessed 10 February 2012.
UNFCC (United Nations Framework Convention on Climate Change) (2007), *Impacts, Vulnerabilities and Adaptation in Developing Countries*, Bonn: UNFCC.
WEF (World Economic Forum) (2011), 'Global competitiveness report 2011–2012', Geneva: WEF, http://www3.weforum.org/docs/WEF_GCR_Report_2011-12.pdf.
World Bank (2007), *Potential and Prospects for Regional Energy Trade in the South Asia Region*, South Asia region, World Bank.

World Bank (2008), *World Development Report: Agriculture for Development, 2008*, Washington, DC: World Bank.
World Bank (2010), 'World development indicators', Washington, DC, http://data.worldbank.org/data-catalog/world-development-indicators.
World Bank (2011), 'World development indicators 2011', Washington, DC, http://data.worldbank.org/data-catalog/world-development-indicators.
World Bank and PPIAF (Public–Private Infrastructure Advisory Facility) (various years), PPI (Private Participation in Infrastructure) Database, http://www.ppiaf.org/ppiaf/page/knowledge-center/private-participation-infrastructure-database.
World Resources Institute (WRI) (2009), 'Energy and resources database', http://earthtrends.wri.org/searchable_db/index.php?theme=6.

PART II

Responsible Policy Matters

4. Taxing for the future: an intergenerational perspective

Tapan Sarker

TAXING FOR THE FUTURE

The European debt crisis has raised fears that fiscal unsustainability in the region is triggering a collapse in the Euro zone, which could pull the European Union, the United States and Japan into recession because of the huge debt levels of these three major Organisation for Economic Co-operation and Development (OECD) regions (OECD 2011). These and other recent fiscal instability issues in Greece, Portugal and Spain have reminded us of the urgent need for fiscal consolidation and for policies to achieve fiscal sustainability that can potentially avoid the world plunging into a second global recession. This is particularly urgent for debt-stricken nations[1] in both the East and the West.

Although raising taxes is one path towards restoring fiscal sustainability, policymakers now, more than ever, need to walk a fine line between economic stimulus and fiscal consolidation, so as to reduce the government deficits without extinguishing domestic economic recovery (Bilicka et al. 2011; McKibbin and Stoeckel 2011). The perils of imposing severe austerity measures were starkly shown in Greece and Italy recently, where serious social and political turmoil resulted, including the resignation of both countries' presidents (Cabannes 2011). There is also growing recognition that the current global patterns of consumption and investment are unsustainable, meaning that many nations must rebalance their economies towards repairing, maintaining and perhaps gaining fiscal sustainability.[2] This is particularly true for many developed economies such as Japan and Republic of Korea, which are experiencing the additional pressures of rapidly ageing populations and sluggish economic growth (Ernst and Young 2011). According to analysts and scholars, reducing their budget deficits through promoting fiscal discipline is one important action. For resource rich economies, restoring fiscal sustainability means that such countries will need to utilize their natural resource endowments to advance social, economic and environmental development outcomes by further broadening the tax base, as well as aggressively enforcing existing taxes (Sarker 2011; Ernst and Young 2011; Scholl et al. 2010; Blanchard and Milesi-Ferreti 2009).

Excessive government debt can also have serious intergenerational consequences, because the debt incurred by today's generations must be paid for by future generations (see, for example, Bowen et al. 1960; Jensen et al. 1999).[3] Auerbach (2008) argues that, since national debt can be viewed as an obligation passed from current generations to future generations:

> . . . there are at least three important long-term objectives that appear to be associated with concerns about debt and deficits: intergenerational equity, economic performance, and fiscal sustainability. Unacceptably high levels of debt are seen as compromising all three objectives. (p. 2)

Effective taxation policies directed at reducing government debt can also have long-term benefits and redistributive impacts (Usui 2011). In the current economic climate the importance of this issue is being increasingly recognized by a range of different bodies including the OECD and International Monetary Fund (IMF). While the literature on taxation is often focused on short-run efficiency and equity considerations, it has given little consideration to how taxation for intergenerational equity, or a 'just tax system' can be practically achieved. It is worth mentioning that arguments about the proper tax baseline commonly concern either: (a) the most appropriate concept of net income, or (b) the choice between an income tax or consumption tax baseline. We note, however, that the difference between an income and consumption tax – the tax burden on capital income – is at the root of many of the complaints about the inefficiency of an income tax (Goode 1977). For instance, income-related benefits are a much more efficient way of increasing the disposable income of poorer households than reduced rates of value added tax (VAT). On the other hand, consumption taxes such as VAT may be the only way to finance (more) strongly progressive spending. However, as some countries lack the administrative capacity to make welfare transfers to households, there may be a case for differentiation in VAT rate structure to tax 'necessities' at a lower rate (OECD 2012).

In order to provide some guidance on how policymakers can promote fiscal sustainability, the following issues will be covered:

- The role of fiscal sustainability to enhance intergenerational equity and thereby long-term sustainable economic development;
- The importance of designing a 'just tax system' for intergenerational justice; and,

- Case studies from both developed (Australia) and developing economies (Bangladesh) in which policymakers are trying to implement measures to improve fiscal sustainability for sustainable economic development.

FISCAL SUSTAINABILITY AND SUSTAINABLE ECONOMIC DEVELOPMENT: A THEORETICAL PERSPECTIVE

The debt crisis, following on the heels of the global financial crisis (GFC) in 2008, demonstrates the negative short and long-term consequences that enduring fiscal unsustainability can have on domestic and global welfare. Cliffe (2011) posits that at least in the short term, fiscal consolidation and reduction of government debt is necessary to avert debt defaults, and the resultant economic contagion and collapse.

There is no doubt that a requirement for intergenerational justice constitutes one of sustainable development's key components. In fact, its most popular definition is development that 'meets the needs of the present without compromising the ability of future generations to meet their own needs' (WCED 1987, p. 53). Consider the famous Native American saying:

> Treat the Earth well: it was not given to you by your parents, it was loaned to you by your children. We do not inherit the Earth from our Ancestors, we borrow it from our children. It refers to a loan contract, the next generation being the lender and the current one the borrower. This is not the only existing proposal. (Gosseries 2008, p. 15)

Although the long-term implications of excessive levels of government debt and government spending have important inter-generational consequences, they generally receive little attention from policymakers, who are potently focused on short-term political gains. The current fiscal unsustainability means that future generations of taxpayers will face an unmanageable burden for something they do not benefit from due to the excessive consumption by the current generation.[4] Furthermore, fiscal unsustainability means that the enduring well-being and quality of life of future generations has been compromised by current generations.[5] Hence, promoting fiscal sustainability is an important component of enhancing intergenerational equity, to prevent future generations bearing the burden of debt incurred by current generations. As argued by Coombs and Dollery (2002):

> Governments can achieve fiscal sustainability simply by raising taxes. . .it is conceivable that a government can address its own financial affairs at the

> expense of its citizens' welfare. Accordingly, in order to ensure that fiscal sustainability is achieved without damage to the welfare of the economy, the concept of optimal taxation should be considered. (p. 12)[6]

The theory of optimal taxation is governed by the Principles of Taxation, which were first developed by Adam Smith in 1776 (Smith 1776). These Principles state that a good tax should be proportional or equitable, certain, simple, transparent, and efficient.[7] When considering the question of intergenerational justice, the equity principle in taxation would at first glance seem to be the most applicable. Equity in taxation is usually assessed by asking who bears the burden of the tax, defined as the change in the distribution of private real income produced by any given tax (see, for example, Kotlikoff and Summers 1987; Sarker 2007). A tax burden that falls most heavily on the poor is said to be 'regressive', while a tax burden that falls most heavily on the rich is said to be 'progressive'. Progressive taxation may often be considered the most socially desirable outcome, since redistributive taxation can subsidize people with low incomes and provide a minimum level of consumption for all (Mirrlees 2011). However, there is a well-acknowledged trade-off between equity and efficiency which policymakers must negotiate, since a steeply progressive tax system can remove incentives to work, save and earn more income (Diamond and Saez 2011; Pressman 2002). As a result, policymakers will often assess whether a taxpayer ought to be bearing that particular tax burden by reference to two principles: the benefit principle, and the ability to pay principle. The benefit principle states that people should pay tax according to the benefits they receive from government spending of that tax revenue. In a resource rich economy such as Australia this would seem to imply that the present generation, which is receiving the benefits of the mining boom, should pay taxes according to those benefits received (Australian Government 2010a). By contrast, the ability to pay principle states that those who are more capable of bearing the burden of taxation should pay more taxes than those with less ability to pay (Auerbach 1983).[8]

However, equity considerations under the Principles of Taxation are focused on determining the short-term redistributive impact of a proposed change in taxation. There are a number of different types of equity that could also be considered, such as inter-temporal equity (impact on long-term decisions); intergenerational equity (future generations); and spatial equity (urban-rural divide) (Australian Government 2010).

It should be noted that the tax expenditures for social security measure a tax benefit relative to a baseline that is somewhere between a comprehensive income tax and a consumption tax. The properly measured tax expenditure relative to a consumption tax baseline would include only those social security benefits that are accorded treatment more favourable than that implied by a consumption tax, which would correspond to including 50 per cent of social security benefits in the recipient's tax base, because consumption tax treatment would allow the exclusion of about one half of the benefits. Thus, the existing tax expenditure is correct conceptually, but is not measured properly relative to a comprehensive consumption tax baseline. Indeed, part of what is measured should be counted as negative tax expenditure relative to a consumption tax baseline because it represents an over-taxation of benefits received (Carroll et al. 2008).

As policymakers are increasingly being required to grapple with designing tax policy to meet long-term challenges, such as intergenerational debt concerns or the ageing population, there is beginning to be a shift away from the traditional concept of equity in taxation to a much broader notion of 'justice' in taxation that takes long-term intergenerational considerations into account.

DESIGNING A JUST TAX SYSTEM

While assessments of equity in taxation look at short-term considerations, notions of a 'just tax system' adopt a longer term, intergenerational perspective. According to Thompson (2002), three different conceptions of societal justice are commonly used: the mutual advantage theory, the entitlement theory, and justice as fairness. The mutual advantage notion of justice is essentially a cost-benefit analysis, and will be satisfied where each person derives the maximum possible gains from voluntary cooperation. The entitlement theory notion of justice focuses on rewards and punishments, and will be satisfied when each person in society gets what they deserve. However, it is the notion of justice as fairness that is most relevant to a tax system, which focuses on the intergenerational distribution of wealth, or intergenerational equity. Justice as fairness holds that all individuals should be able to enjoy a fair share of resources regardless of their history, opportunities, or deserts (Thompson 2002). Under the principle of justice as fairness, current and future generations should have the same access to resources, wealth and consumption opportunities, and intergenerational rebalancing may be required where this does not currently occur (Ablett 1996). In addition, savings occur when one generation transfers to the next, a capital (in the extended meaning of the word) that is greater than the one it inherited from the previous generation. Inversely, there are generational dissavings whenever one generation transfers to the next a capital that is smaller than the one it inherited (Daly 1996).

In order to move beyond a merely theoretical notion of intergenerational rebalancing, it is necessary to be able to quantify the extent of generational imbalance to determine the adjustment required to rebalance the generations. Generational accounting is one tool that can be used to determine the amount of intergenerational rebalancing necessary to achieve 'justice'. It is based on the premise that future generations should not pay a higher share of their lifetime incomes to the government than current generations do (Auerbach et al. 1994).

For those who associate justice with reciprocity, indirect reciprocity is quite a potent idea. It has the advantage of justifying obligations to people who so far have never given us anything and who may be giving us less in the future than what we will have given them. In the case of direct reciprocity, it is the original benefactor who ends up getting back what he put in, whereas with indirect reciprocity, there is a third party who benefits – in this case, the next generation – instead of the initial benefactor – the previous generation – which gives rise in this way to a chain of obligations (Gosseries 2008). In any event, what really matters in this context, is to emphasize the need to check whether indirect reciprocity really reflects our intuitions about justice, both intergenerationally and as a component of a general theory of justice. The simplest method for such a purpose consists in testing the idea of reciprocity in an intergenerational context. Take for example the case of a person with multiple congenital disabilities. Let us accept the idea that she will give us less in return for what we as a society gave to her – which is not meant to deny the benefits we may of course derive from her company. Given such an example, the limitations of the idea of reciprocity are clear (Fiekowsky 1980).

Rawls in his masterpiece *A Theory of Justice* (1999) is aware of utilitarianism's major difficulties in the intergenerational context. At the same time, he considers that moving away, from the initial condition of prehistoric men is necessary, not just for reasons of efficiency, but for reasons of justice. How can both these concerns be accommodated? By defending a 'two-stage' model in which a steady state phase follows an accumulation phase. During the accumulation phase, principles are identical to those of utilitarianism (compulsory savings). But this phase is supposed to have a limited duration. And the rationale underlying the need for such accumulation is totally unrelated with maximizing the size of the intergenerational welfare pie (Rawls 1999).

The aim of the accumulation phase, Rawls adds, is to allow economic affluence to build up so that at least minimal stability to just institutions can be ensured. As soon as this point is reached, accumulation ceases to be an obligation and the steady state phase begins. And for that second phase, the principle defended by Rawls is identical to the one defended by the indirect reciprocity view (Gosseries 2008). In contrast to the above formulation, Daly argues that '. . . the basic needs of the present should always take precedence over the basic needs of the future but the basic needs of the future should take precedence over the extravagant luxury of the present' (Daly 1996, p. 36). In other words, the most appropriate interpretation is that of 'basic needs'. But even with a less restrictive version, there is still in such a definition the idea that once everyone's needs are covered, fairness does not require any further redistribution (Gosseries 2008).

Generational accounting states the government's intertemporal budget constraint as the present value of remaining net tax payments of existing generations plus the present value of net tax payments of future generations equals the present value of all future government consumption less government net wealth. Generational balance, therefore, is assessed by comparing the per capita fiscal burdens of current and future generations. If the net fiscal burden of the future generation is significantly higher than the current generation, the assumed policy scenario is unsustainable (Auerbach et al. 1994; Ablett 1996; Coombs and Dollery 2002). The generational accounting framework described above shows that an increase in the consumption of current generations, without an increase in the tax currently paid, will require an increased fiscal burden to be borne by future generations (Auerbach et al. 1994; Deutsche Bundesbank 2001). Generational accounting can thus be used as a tool to design policies aimed at achieving intergenerational equity.

Norway's sovereign wealth fund (SWF) is a leading example of a policy designed to achieve intergenerational equity. According to the United Nations' Human Development Index, Norway is the country with the highest living standards in the world (UN 2011). The country owes its prosperity to exporting its rich oil and gas deposits. However, resource-rich countries typically spend their resource wealth, which means current generations will experience higher standards of living than future generations (Corden and Neary 1982). By contrast, Norway has adopted a different approach. Norway's SWF was established to retain and invest surpluses from taxing the country's oil and gas sales at a high rate (Sunley et al. 2002). The SWF will maintain the living standards and pension payments of future generations when its natural resource deposits are depleted. Norway's SWF is now the second-largest SWF in the world, and is an excellent example of how policymakers can secure intergenerational equity by taxing current generations to provide for future generations (Reiche 2008; Norwegian Ministry of Finance 2007; Kunzel et al. 2011), while also facilitating the transition from one form of capital to another.

The Five Capital Model is another example that can be used to provide a framework for examining trade-offs between the different but interconnected forms of capital, which are: natural capital, human capital, social capital, manufactured capital, and financial capital (Brereton and Pattenden 2007; Sigma Project 2003). The Five Capitals framework implies that these forms of capital should be repaired, maintained and enhanced rather than depleted or degraded (Mikesell 1989). Since taxation policy can impact business incentives, it can be used to facilitate intergenerational redistribution by assisting the transition from one form of capital to another. Rather than depleting one form of capital altogether (Mikesell 1989), the focus should be on achieving optimal outcomes across all forms of capital (Southalan 2011). Thus, the Five Capital model implies that the income generated from natural resource extraction (that is, depletion of natural capital) should not be used for consumption. Instead, the revenue from natural resources should be retained and used to develop another form of capital, such as human or social capital. Norway's SWF is an example of this: part of the revenues from oil and gas extraction (the depletion of natural capital) are retained and used for pension payments for a future ageing population (that is, enhancing human capital). This also applies to developing countries. Collier (2010) argues that

> in resource-rich low-income countries . . . the key fiscal issue to be addressed is the replacement of depleting natural assets with other assets, real and financial . . . the revenues need to be earmarked for investment. (p. 78)

Figure 4.1 graphically shows how the components of a just tax system can be derived from the above discussion. A just tax system should achieve fiscal sustainability and intergenerational equity; it should maintain and develop, rather than deplete capital; and it should be focused on achieving economic growth that is sustainable in the long run, rather than being focused on short-term gains.

The leftmost circle speaks of the relationship between intergenerational equity and fiscal sustainability. Since national debt can be inherited by future generations, fiscal sustainability is necessary to reduce excessive national debt and achieve intergenerational balance. The circle on the right describes how the Five Capital Model advocates capital maintenance rather than depletion. Fiscal sustainability for the long term often requires investing in the five forms of capital and ensuring it is maintained for future generations. This capital investment may take a range of forms, including in education and human capital development, or Norway's SWF. Finally, sustainable economic development speaks of economic growth that is sustainable over the long term, and which takes into account social and

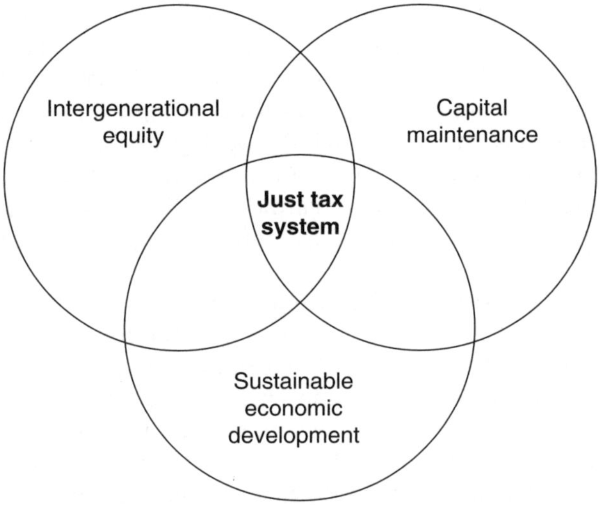

Figure 4.1 A just tax system

environmental welfare. Fiscal sustainability and intergenerational equity are necessary to achieve government spending and economic growth that is sustainable in the long run. Capital maintenance will also aid in achieving economic growth that is sustainable in the long term, since extensive capital depletion will eventually hinder economic growth. As shown in Figure 4.1, all three of these long-term policy objectives are interrelated, and it is this nexus of the three that can be used to describe a 'just' tax system that simultaneously achieves all three objectives.

CASE STUDIES

Policymakers all over the world are now trying to find ways towards achieving fiscal sustainability through effective fiscal policy reforms. While the issue is most pressing in debt stricken developed countries such as the USA and Japan, developing countries are also trying to create some drastic policy changes to address long-term social and economic challenges by redesigning their taxation systems.

Because there are differences in what income can mean, there is substantial disagreement over the appropriate baseline even among those who accept the idea that an income tax is the (or perhaps *an*) appropriate baseline against which to identify and measure tax expenditures. The choices typically discussed include comprehensive income as well as income concepts that lie somewhere between current law's tax base and comprehensive income. In addition, comprehensive income (sometimes called Haig-Simons income or economic income) is defined as accretion to an individual's real purchasing power occurring within a year, that is, as consumption plus the change in inflation adjusted net worth or saving. It is generally considered the 'ideal' or 'pure' base for an income tax. It includes all accretions to wealth, regardless of whether realized, related to a market transaction, or representing a return to capital or labour (Goode 1977).

Furthermore, a criticism of consumption tax is that it is not always clear how to treat certain items under a consumption tax. One problem is determining whether a particular expenditure is an item of consumption. Spending on medical care and charitable donations are two such examples. Furthermore, there may be more than one way to treat various items under a consumption tax. For example, a consumption tax might ignore borrowing and lending by excluding from the borrower's tax base the proceeds from loans, denying the borrower a deduction for payments of interest and principal, and excluding interest and principal payments received from the lender's tax base (Carroll et al. 2008).

In this section, I will present two case studies, representing developed and developing countries (Australia and Bangladesh, respectively), where policymakers are trying to implement policy measures to improve fiscal sustainability to achieve sustainable economic development.

Australia

Standing still is not an option for the millions of Australians whose future is at stake. Australia is a developed nation with the world's largest resources of recoverable brown coal, lead, nickel, tantalum, uranium and zinc (DFAT 2008). It is for this reason that earnings from natural resource exports amount to almost half of Australia's total export earnings, making it the country's largest single export sector (DFAT 2008). However, concerns have been raised about the danger of relying solely on finite natural resource deposits to drive Australia's economic growth, particularly where that comes at the expense of other sectors. In Australia, for example, booming natural resource exports have appreciated the Australian dollar, and adversely affected the international competitiveness of other sectors, such as manufacturing and agriculture. This problem is compounded by the fact that most of the spillover benefits of mining go overseas (RBA 2011). Further, most countries collect a minimum stable revenue stream from mining operations, while booming commodity prices lead to the state retaining an ever-smaller proportion of natural resource revenues

(Australian Government 2010a; Southalan 2011). Finally, mining is a cyclical industry, and spending natural resource revenues on consumption, rather than productive investment, will result in lower standards of living for future generations, compared to current standards, when natural resource deposits are exhausted (Corden and Neary 1982).[9] This is a common problem. McPherson (2010, p. 280) provides an explanation of the concept of the 'natural resource curse', which applies equally to mineral extraction:

> Due to the dramatic increase in oil prices, most oil producing countries have been able to record overall budget surpluses. At the same time many are significantly increasing the size of their non-oil spending and non-oil deficits, exposing several of them to serious fiscal risks should the oil price drop sharply.

It is for this reason that the Australian Government recently proposed a tax be imposed on the profits of mining companies (Australian Government 2010b; Sarker and Whalan 2010) to mitigate the effect of the mining boom and help Australia for the transition out of its 'two-speed economy' (Dutch disease). A tax on the supernormal profits generated by natural resource extraction has attracted an increasing amount of attention as an efficient way 'to tax the exploitation of a heterogeneous resource base in conditions of economic uncertainty' (Land 2010). The Australian Government proposed that revenue from the tax be used to finance infrastructure spending, or invested in the wealth of future generations (AIG 2010). For political reasons, this proposal went ahead only in a significantly modified form (Sarker and Whalan 2010; Southalan 2011).

Increasing the taxation of the mining sector is an example of 'just' taxation that focuses on the long-term sustainability of the economy, since it can help an economy avoid the Dutch Disease syndrome (Corden and Neary 1982). This describes a situation in which resource-rich economies forget that natural resource deposits are finite and engage in irresponsible spending of windfall resource revenues while they exist, rather than investing the money to benefit future generations. This may also cause weak political institutions, rent seeking, excessive borrowing, wasteful spending, unsustainable economic growth, and a 'two-speed' economy (where the mining sector is booming, but other longer-term industries are suffering as a result). By contrast, investment of natural resource revenues can secure a country's economic growth beyond the mining boom (Kumar 1991). Collier (2010), for example, argues that the appropriate use of natural assets is to accelerate the evolution towards the eventual level of sustainable income. Productive uses for natural resource wealth may include public capital investment, reduction of public debt, and retaining wealth for future generations (like Norway's sovereign wealth fund).[10] Redirection of natural resource wealth to these more productive uses requires collection of this revenue through the taxation system. These taxation revenues can also be used to alleviate the negative impacts of mining on local communities and the environment (Mikesell 1989; Andrews-Speed and Rogers 1999; Clark and Clark 1999), in addition to important redistributive outcomes (Berlinschi and Daubanes 2010). Hence a profits-based mining tax can help secure intergenerational equity, and also lead to significant efficiency gains (Garnaut and Clunies-Ross 1983). Other international examples of where a resource profit-based tax has been introduced include Malawi, which applies a 10 per cent resource rent tax (RRT) above a certain threshold; Norway, which approximates a Brown tax through its corporate income tax system, and Mongolia, which has a windfall profits tax based on commodity price rises (Hogan and Goldsworthy 2010).

The question of an appropriate policy response to a rapidly changing climate has also been at the centre of the Australian political debate in recent times (Combet 2011), particularly since climate change has significant intergenerational implications. Through a rapidly changing climate, extreme weather conditions and other serious negative effects will change the world as we know it (Parry et al. 2007). The pollution of today's generation will impose large costs on future generations, including increased pressures on the natural environment and reduced opportunities for development.[11] Therefore, achieving intergenerational equity means taking action to reduce pollution now to reduce the costs imposed on future generations: an equitable division of emission rights (Stanton and Ackerman 2009). Environmental taxation, in various forms, is often suggested as a way to restore intergenerational balance by requiring large polluters, such as the energy sector, to pay for the costs their pollution imposes on others, thus creating an incentive for large polluters to reduce the amount of pollution.

It is for this reason that a number of countries have imposed a carbon tax, or provided other tax incentives for the development of low-carbon sustainable energy sources. Thirty-two countries, including the European Union, and ten US states already have a carbon price that is set by an emissions trading scheme. China, Taiwan, Chile, Republic of Korea and a number of Canadian provinces are also considering a carbon-trading scheme, or already have trial schemes in place (Garnaut 2011). It is for these reasons that Australia followed suit in 2011 (Australian Government 2011). A carbon trading scheme acts to restore intergenerational balance by imposing some of the costs of climate change on current generations and reducing the costs of climate change that future generations will bear, and is accordingly another example of taxing for the future.

Another problem of intergenerational justice that needs to be urgently addressed is that of the ageing population. This is a crucial issue for governments across the world, with the prospect of a rapidly ageing population causing low economic growth, due to higher government pension liabilities and a lower proportion of the population working and paying taxes (AIG 2010; Takayama and Kitamura 1999). The growing ranks of elderly (with ever-longer life expectancies) are expected to place significant pressure on health care, pensions, infrastructure, productivity and GDP (Swan 2010; Henry 2009), since future productivity growth is not expected to fully offset these costs (Henry 2009; cf McDonald 2005). This raises several issues of intergenerational justice. First is the uneven level of projected public spending: if a government is unable to fund pension liabilities and health care services, then future generations of elderly experience much lower living standards than current generations (Coombs and Dollery 2004). Secondly, this ageing population results in intergenerational inequity: the young have to pay for the living standards of the old, and these current welfare benefits will not be present when young taxpayers become elderly (Rydell 2005). Redressing the generational imbalance means increasing tax revenues on current generations or increasing public borrowing (Agulnik et al. 2000; Cortese and Glynn 2005). Many argue, from an intergenerational perspective, that current public and private saving should be increased (through increased taxation and superannuation contributions) and welfare should be cut in order to redress the intergenerational imbalance (Rydell 2005; Coombs and Dollery 2004).

The Australian Government has adopted a three-pillar approach to the ageing population: a publicly provided age pension, compulsory saving through the superannuation scheme, and voluntary superannuation saving. The Australian Government has also recently reviewed the question of the ageing population and intergenerational equity in some detail (Henry 2009; Harmer 2009), and concluded that the current age pension rate is generally adequate (Harmer 2009), and that the current superannuation guarantee rate of 9 per cent is adequate to provide most people with substantial income replacement. However, the age pension minimum age should be increased in line with increasing life expectancies (Henry 2009). Dealing appropriately with the question of the ageing population is a critical requirement for securing intergenerational justice, and Australia, with a compulsory 9 per cent superannuation scheme since 1992, has long known the importance of taxing for the future. It is because of this foresight that the country is likely to fare a lot better than the similarly developed Asian nations of Japan and Korea (see, for example, Takayama and Kitamura 1999).

As one of the most developed nations in Asia, Australia is facing a number of pressing intergenerational challenges, including: avoiding the Dutch disease and preserving the mining wealth for future generations; finding an appropriate response to climate change; and responding to the ageing population. Interestingly, according to recent surveys, polls consistently show that around 9 in every 10 Australians want more renewable energy, but this will only happen with well designed and efficient government support that allows businesses to plan and grow and reduce their cost until they are competitive with other forms of energy. We now have a 20 per cent Renewable Energy Target supported by all major political parties, and a price on carbon that started on 1 July 2012. These investment-grade policies will form the centrepiece of a national approach to reduce carbon emissions and help clean energy technology become fully competitive with mainstream energy (Thornton 2012). The examples discussed above show how well thought sound fiscal policy could assist policymakers to achieve intergenerational justice, by 'taxing for the future' and collecting revenue of current economic activities to provide better support for long-term problems.

Bangladesh

Bangladesh is a developing nation located in South Asia, with an estimated 40 per cent of the population living below the poverty line (World Bank 2011). Given the extremity of the development problems experienced by Bangladesh, policies that work these are likely to provide guidance to other developing economies experiencing less extreme conditions. Further, it is important to realize that taxes that work in developed countries may not be effective in a developing country context (see, for example, Mirrlees 2011; Tanzi and Zee 2001). According to ADB (2004) and others (Kim 2005; IMF 2002), effective fiscal policies directed at achieving sustainable economic development are crucial to poverty alleviation in the developing countries such as Bangladesh. The country is also highly vulnerable to the impacts of climate change due to its location at the delta of a number of important rivers ending in the Bay of Bengal, and the large proportion of low-lying land subject to inundation as sea levels rise (Sarker et al. 2012). This vulnerability to climate change poses a serious impediment to the long-term economic development of Bangladesh (Agrawala et al. 2003).

Apart from the climate change issue, Bangladesh also faces a mounting task to increase funding for education and human capital development that can help long-term economic development.[12] Bangladesh's new education policy has addressed this issue and has notably received much attention in recent years as a key means of improving the nation's GDP growth and redressing inequality. It is estimated that between 42 per cent and 51 per cent of the population is currently illiterate, and 38 per cent of the population has received no formal education at all (Ministry of Education 2010; BBS 2010). The recent National Education Policy 2010 of Bangladesh shows recognition of the importance of expanding the provision of quality education in this nation, which according to education policy experts will require more than Tk 30 000 crore (approximately $37.5 billion AUD) (Haque 2011). However, state-provided or subsidized education raises the question of funding, with developing countries in particular facing severe funding limitations (IIEP 2007). Developing countries seeking to alleviate poverty by expanding the education sector are facing acute financing problems, and hence must look to new sources of revenue, mainly through internal resource mobilization, which requires effective fiscal policy reform (Lewin and Caillods 2001; Wang et al. 2011). This is true for Bangladesh, which faces huge funding costs for its ambitious education plan.

However, finding a new tax base to fund and implement such an important policy is a daunting task for any developing country. One possible option suggested was expanding the tax base on land and property, which also has received renewed attention in recent years. Importantly, a degree of transparency is provided since the property tax would be imposed on a relatively immobile base (making it difficult to evade), and is based on land and property valuations (FAO 2002). Education and tax experts of Bangladesh have suggested that a property tax could address the problem of education funding and raise substantial revenue, if managed properly (Sarker 2011; Haque 2011). This taxation method has been used in other countries including the USA to fund local schools (Kenyon 2007). One way this type of taxation could be implemented is if revenue collected from such taxes is separately allocated for the purpose of funding local schools. This may have significant redistributive impact by increasing the taxation on the wealthier members of society, and contributing funds to raising the education level, which is crucial for poverty alleviation and economic growth (Sarker 2011; Haque 2011).

Increasing a country's education has significant short-term and long-term benefits at both the individual and social levels. Education is a well-recognized driving factor behind increased social mobility, reduced inequality, increased living standards and poverty alleviation. According to a number of international bodies (IFC 2001; OECD 2010; World Bank 2011), it also increases the productivity of the labour force, increases health, increases economic growth, drives economic competitiveness, and promotes democracy, peace, stability, and environmental preservation. Education is thus not only crucial to achieving short-term economic growth, but also to achieving long-term intergenerational equity objectives such as higher living standards and lower poverty rates. A property tax in Bangladesh may assist in achieving these goals through easing the inequality as well as investing in education, which can provide a double dividend for such a poor nation.

A second tax in Bangladesh that may assist in achieving intergenerational equity is VAT (see, for example, Faridy and Sarker 2011; Sarker 2007). Value added tax is widely considered to be one of the more efficient taxes, the introduction of which in place of more narrowly based indirect taxes can lead to significant efficiency gains (Ebrill et al. 2002; KPMG 2011). OCED member countries have relied increasingly on VAT as a source of revenues. It has become the third largest source of revenue for OECD countries as a whole, behind personal income taxes and social security contributions. Its neutrality principle towards international trade has also made it the preferred alternative to customs duties and sales taxes in the context of trade liberalization (OECD 2012). Complex, unclear or inconsistent rules across jurisdictions are difficult to manage for revenue bodies and create uncertainties and high compliance costs, which can lead to reduced compliance. Such an environment may also facilitate tax fraud and avoidance.[13]

Bangladesh introduced a VAT in 1991, but its many exemptions have hindered its revenue raising potential (Faridy and Sarker 2011), thus there is a strong argument for broadening the base of Bangladesh's VAT. Effective mobilization of internal revenue is crucial to the ongoing economic development of Bangladesh, since government-provided welfare can assist in achieving poverty alleviation. Mobilization of internal revenue sources can also help to ensure the long-term fiscal sustainability of Bangladesh, since it can help reduce the country's dependence on foreign funding sources, and therefore reduce government deficits (Faridy and Sarker 2011). Similar arguments have been made in the United Kingdom in the context of the economic recovery from the GFC, where it was proposed that the VAT rate be increased to 20 per cent to assist the government's fiscal consolidation and debt reduction (Fuest 2009). Obtaining enough foreign loans to finance its $395 billion BDT (Bangladeshi takas) deficit has been a problem for Bangladesh in the past (Xinhua 2010). Essentially, broadening the VAT base will help Bangladesh achieve fiscal sustainability, which helps intergenerational rebalancing.

Achieving intergenerational justice is an important issue for both developed and developing nations in Asia. Well-thought out taxation policy focusing on long-term considerations can assist countries in dealing with these long-term challenges. It is evident from the above examples that justice in taxation requires achieving intergenerational balance. Developed nations, such as Australia, will usually be required to use taxation revenue to provide for future costs that will arise, such as the reduction of government debt, climate change, the ageing population, or the cessation of the mining boom. Developing nations such as Bangladesh will usually use their taxation revenue to increase the long-term economic growth of a nation through health, education, infrastructure, or reduction of government debt. All of these long term development and equity objectives require a taxation system that is focused on the long term: taxation for the future.

CONCLUSION

This chapter discusses the concept of taxing for the future and provides a discussion of the traditional principles of taxation and why they are inadequate to deal with long-term considerations of intergenerational justice. It provides an analysis of 'justice' in taxation and suggests a new paradigm for designing tax policy that incorporates considerations of fiscal sustainability, capital maintenance and sustainable economic development. Without fiscal sustainability, no economic development strategy can succeed. Unfortunately, in many emerging markets and developing countries, weak political institutions and incompetent and corrupt public administrations impose very strict limits on the amount of debt, internal and external, that can be carried. This new approach to fiscal policy ensures revenue raising that is sustainable (present into future) and can bring long-term sustainable economic development.

First, it discusses fiscal sustainability issues and how to achieve fiscal sustainability to enhance intergenerational justice. The study finds that excessive government spending by the current generation will require future generations to bear the burden of this spending since debt incurred by current generations will be inherited by future generations. Second, this chapter focuses on how the principles of taxation, particularly the concept of equity in taxation, need to be broadened to take into account long-term intergenerational considerations. The chapter then considers what the components of a 'just' tax system are and how justice can be measured. It also looks at how generational accounting can determine the extent of generational rebalancing, and the way the Five Capital Model can be used to give direction to sustainable fiscal policy by aiding in capital maintenance rather than depletion. A just tax system is achieved at the nexus of fiscal sustainability, capital maintenance and sustainable economic development. Finally, this chapter looks at how developed and developing countries have implemented the concept of a 'just' tax system by taking the case studies of Australia, a developed nation in Asia, and Bangladesh, a developing nation in Asia.

Achieving fiscal sustainability will help an economy achieve long-term and sustainable development that balances economic, social, and environmental considerations. Fiscal sustainability requires responsible fiscal policy, which necessitates taxing and spending with regard to intergenerational considerations. Good fiscal policy must be implemented with good fiscal instruments that assist the maintenance, rather than the depletion of capital. These may include: establishing a sovereign wealth fund or future fund for the next generation; taxing economic 'bads' such as mining or pollution; lowering income tax and corporate tax to increase production incentives and replacing them with broader-based efficient taxes such as the VAT; and efficient revenue raising for capital development, such as a property tax to fund educational development.

The following considerations for policymakers designing a just tax system can be drawn from the above discussion:

- the long-term social, environmental and economic sustainable development challenges faced by the country;
- the long-term distributional implications of the current taxation system;
- easing the intergenerational imbalance (shown by generational accounting);
- the optimal level of debt borne by future generations;
- capital maintenance rather than depletion;
- taxing unsustainable 'bads' to incentivize development that is sustainable;
- reallocating funds to increase long-run economic productivity and sustainable development;
- broadening the tax base and moving to more efficient taxes;
- considering long-term issues (ageing population, climate change) in present-day policy development and;
- efficient, simple and transparent fiscal policy implementation.

NOTES

1. Countries with very high debt to GDP ratios include Japan at 471 per cent (although partially offset by high levels of household saving); the United Kingdom at 466 per cent; Spain at 366 per cent; South Korea at 333 per cent; France at 323 per cent; Italy at 315

per cent; the United States at 296 per cent; and India at 129 per cent (McKinsey Global Institute 2010).

2. For a comprehensive review, please see IMF (2011).
3. This means that the debt reduction policies of today's government will have an impact on the level of intergenerational equity of the future generations particularly in their living standards and distribution of public resources (Australian Government 2002).
4. Thompson (2002) shows that, under the benefit principle, those who do not benefit from something should not be required to pay for it.
5. This is particularly important in the context of the ageing population, where many governments are faced with the reality of being unable to provide an age pension for the growing numbers of retirees they will encounter (Thompson 2002).
6. Hence, effective fiscal reform and policy instruments can enhance, maintain or repair fiscal sustainability, thus playing a role in helping a country achieve sustainable economic development.
7. For a fuller explanation of the principles of optimal taxation, see, for example, Musgrave and Musgrave (1989), Slemrod and Bakija (2008).
8. The principle of equity also implies both horizontal and vertical equity. Horizontal equity states that individuals in similar economic positions should be treated alike. Vertical equity holds that those in different economic positions should be treated differently. These principles usually imply a progressive tax system, where those who have greater economic capacity will pay more (Australian Government 2010). Policymakers will often be required to resolve the conflict between equity and efficiency.
9. See, for example, Weinthal and Luong (2001); Sandbu (2006); Southalan (2011); Van der Plog and Venables (2011).
10. See, for example, Weiss (2008); Sandbu (2006); Van der Plog and Venables (2011); Weinthal and Luong (2001).
11. See, for example, Brodhag (2010); Stanton and Ackerman (2009); Liu (2010); Soltau (2006).
12. Education funding is crucial for future growth and development (UNESCO 2008). History suggests that the funding in education and innovation and science are central for future development, as evidenced by recent experience in China, for which expenditure on state funded education played a crucial role in alleviating poverty in China from 250 million in 1978 to 28.2 million in 2002 (UNESCO 2008; Ghosh 2010).
13. Research by OECD's Committee on Fiscal Affairs (CFA) in co-operation with business, has shown that the current international consumption taxes environment, especially with respect to trade in services and intangibles, is creating obstacles to business activity, hindering economic growth and distorting competition.

REFERENCES

Ablett, J. (1996), 'Generational accounting and intergenerational balance', *Agenda*, **3** (4), 407–18.

Agrawala, S., T. Ota, A. Ahmed, J. Smith and M. Van Aalst (2003), 'Development and climate change in Bangladesh: focus on coastal flooding and the Sundarbans', Organisation for Economic Co-operation and Development Report, www.oecd.org/dataoecd/46/55/21055658.pdf.

Agulnik, P., R. Cardarelli and J. Sefton (2000), 'The pensions green paper: a generational accounting perspective', *The Economic Journal*, **110**, 598–610.

AIG (Australian Industry Group) (2010), 'Managing Australia's mining boom', Economics Research Note, Australian Industry Group, http://www.aigroup.com.au/economicsresearch/researchnotes.

Andrews-Speed, P. and C. Rogers (1999), 'Mining taxation issues for the future', *Resources Policy*, **25**, 221–7.

Asian Development Bank (2004), 'Economic growth and poverty reduction in Bangladesh', www.adb.org/Documents/Studies/Eco_Pov_Ban/Eco_Pov_ban.pdf.

Auerbach, A. (1983), *The Taxation of Capital Income*, Cambridge, MA: Harvard University Press.

Auerbach, A. (2008), 'Long-term objectives for government debt', Conference on Fiscal Policy and Labour Market Reforms organized by the Swedish Fiscal Policy Council (Finanspolitiska Rådet), Stockholm, 29 January 2008.

Auerbach, A., J. Gokhale and L. Kotlikoff (1994), 'Generational accounting: a meaningful way to evaluate fiscal policy', *Journal of Economic Perspectives*, **8** (1), 73–94.

Australian Government (2002), 'Budget 2002–2003', Canberra: Department of Treasury, http://www.budget.gov.au/2002-03/bp5/download/03_BP5Part1.rtf.

Australian Government (2010), 'What does equitable distribution mean?', Australia's Future Tax System: Architecture of Australia's Tax and Transfer System, Department of Treasury http://taxreview.treasury.gov.au/content/Paper.aspx?doc=html/publications/papers/report/section_3-02.htm (Henry Tax Review).

Australian Government (2010a) *The Resources Super Profits Tax: A Fair Return to the Nation*, Canberra: Department of Treasury.

Australian Government (2010b), 'Australia's future tax system: final report part 1', Canberra: Department of Treasury, http://www.futuretax.gov.au.

Australian Government (2011), 'Working together for a clean energy future: our plan', http://www.cleanenergyfuture.gov.au/clean-energy-future/our-plan/.

BBS (Bangladesh Bureau of Statistics) (2011), 'Report on the Bangladesh Literacy Survey', Industry and Labour Wing, Dhaka: Bangladesh Bureau of Statistics, Ministry of Planning.

Berlinschi, R. and J. Daubanes (2010), 'Foreign aid and oil taxes: helping the poor in oil-rich countries', http://www.unige.ch/ses/ecopo/seminaires/ecometri/Berlinschi.pdf.

Bilicka, K., M. Devereux and C. Fuest (2011), 'Will fiscal consolidation drive the UK economy back into recession?', Oxford Centre for Business Taxation Policy Briefing 21 March http://www.sbs.ox.ac.uk/centres/tax/Pages/default.aspx.

Blanchard, O. and G. Milesi-Ferreti (2009), 'Global imbalances: in midstream?', International Monetary Fund Staff Position Note No 9, http://www.imf.org/external/np/exr/key/finstab.htm.

Bowen, W., R. Davis and D. Kopf (1960), 'The public debt: a burden on future generations?', *American Economic Review*, **50** (4), 701–6, http://www.jstor.org/stable/1812466.

Brereton, D. and C. Pattenden (2007), 'Measuring what matters: monitoring the contribution of a new mining project to community sustainability', 3rd International Conference on Sustainable Development Indicators in the Minerals Industry (Milos Conferences), Milos, Greece, 18 June, http://www.csrm.uq.edu.au/docs/Brereton%20Pat tenden%202007%20SDIMI.pdfS.

Brodhag, C. (2010), 'A differentiated approach for sustainable consumption and production possibilities', *Natural Resources Forum*, **34**, 63–70.

Cabannes, A. (2011), 'The European monetary crisis explained', SSRN Working Paper Series, 18 August, http://ssrn.com/abstract=1911816.

Carroll, R., D. Joulfaian and J. Mackie (2008), 'Income versus consumption tax baseline for tax expenditures', in *Incentive and Distributional Consequences of Tax Expenditures*, forthcoming, presented March 2008.
Clark, A. and J. Clark (1999), 'The new reality of mineral development: social and cultural issues in Asia and Pacific nations', *Resources Policy*, **25**, 189–96.
Cliffe, M. (2011), 'Losing the euro would spell disaster', *De Volkskrant*, 1 December 2011, http://www.ing.com/Our-Company/About-us/Features-archive/Collapse-of-the-euro-would-have-disastrous-effects.htm.
Collier, P. (2010), 'Principles of resource taxation for low-income countries', in P. Daniel, M. Keen and C. McPherson (eds), *The Taxation of Petroleum and Minerals: Principles, Problems and Practice*, New York: Routledge, pp. 75–86.
Combet, G. (2011), 'Tackling climate change is in the national interest', speech to the Press Club, Canberra, 13 April, http://www.climatechange.gov.au/minister/greg-combet/2011/major-speeches/April/sp20110413.aspx
Coombs, G. and B. Dollery (2002), 'An analysis of the debate on intergenerational equity and fiscal sustainability in Australia', University of New England, Working Paper Series in Economics Discussion Paper No. 5, https://www.une.edu.au/bepp/working-papers/economics/1999-2007/econwp02-5.pdf.
Coombs, G. and B. Dollery (2004), 'The economics of ageing: generational accounting and regional public goods in Australia', University of New England Working Paper Series in Economics Discussion Paper 6, www.une.edu.au/febl/EconStud/wps.htm.
Corden W. and P. Neary (1982), 'Booming sector and de-industrialisation in a small open economy', *The Economic Journal*, **92** (368), 825–48.
Cortese, C. and J. Glynn (2005), *Taxing Our Future: The Burden of the Australian Superannuation Taxation System*, Conference proceedings, 4th Global Conference of Business and Economics, 26–28 June, Oxford: Oxford University.
Daly, H. (1996), *Beyond Growth, The Economics of Sustainable Development*, Boston, MA: Beacon Press.
Deutsche Bundesbank (2001), 'The long-term sustainability of public finance – an analysis based on generational accounting', *Monthly Report of the Deutsche Bundesbank*, **53** (12), 29–43.
DFAT (Department of Foreign Affairs and Trade) (2008), 'About Australia: resources sector', Commonwealth Government of Australia, Canberra, http://www.dfat.gov.au/facts/resources_sector.html.
Diamond, P. and E. Saez (2011), 'The case for a progressive tax: from basic research to policy recommendations', CESifo Working Papers No. 3548, August 2011, http://elsa.berkeley.edu/~saez/diamond-saezJEP11opttax.pdf.
Ebrill, L., M. Keen, J. Bodin and V. Summers (2002), 'The allure of the value-added tax', Finance and Development, International Monetary Fund, http://www.imf.org/external/pubs/ft/fandd/2002/06/ebrill.htm.
Ernst and Young (2011), 'Asia-Pacific tax policy outlook', http://www.ey.com/GL/en/Services/Tax/Business-Tax/Tax-Policy-and-Controversy/The-outlook for-tax-policy-around-the-world.
FAO (Food and Agriculture Organization) (2002), 'Rural property tax systems in Central and Eastern Europe', FAO Land Tenure Studies, http://www.fao.org/DOCREP/005/Y4313E/y4313e05.htm.
Faridy, N. and T. Sarker (2011), 'Progressivity of value added tax (VAT) in developing countries: empirical evidence from Bangladesh', *Asia Pacific Tax Bulletin*, **17** (3), 185–91.

Fiekowsky, S. (1980), 'The relation of tax expenditures to the distribution of the fiscal burden', *Canadian Taxation*, **2** (Winter), 211–19.
Fuest, C. (2009), 'The UK should increase VAT', *The Tax Journal*, 24–25, http://www.sbs.ox.ac.uk/centres/tax/Pages/default.aspx.
Garnaut, R. (2011), 'Carbon pricing and reducing Australia's emissions', Garnaut Climate Change Review Update 2011, Update Paper 6, available at http://www.garnautreview.org.au.
Garnaut, R., and A. Clunies-Ross (1983), *Taxation of Mineral Rents*, Oxford: Oxford University Press.
Gerson, P. (1998), *Poverty, Income Distribution and Economic Policy in the Philippines*, International Monetary Fund Fiscal Affairs Department Working Paper No. 20, Washington, DC: International Monetary Fund.
Ghosh, J. (2010), 'Poverty reduction in China and India: policy implications of recent trends', DESA Working Paper No. 92, United Nations Department of Economic and Social Affairs, http://www.un.org/esa/desa/papers.
Goode, R. (1977), 'The economic definition of income', in Joseph Pechman (ed.), *Comprehensive Income Taxation*, Washington, DC: The Brookings Institution, pp. 1–29.
Gosseries, A. (2008), 'Theories of intergenerational justice: a synopsis', SAPIENS [online] 1.1, online since 23 December 2008, http://sapiens.revues.org/165, accessed 20 February 2013.
Haque, M. (2011), 'Property tax be imposed to finance education policy implementation', *Holiday*, 1 April, http://www.weeklyholiday.net/Homepage/Archive/010411/com.html.
Harmer, J. (2009), 'Pension review report', Australian Government Department of Families, Housing, Community Services and Indigenous Affairs, Canberra: Commonwealth Government, http://www.fahcsia.gov.au/sa/seniors/progserv/PensionReview/Pages/default.aspx.
Henry, K. (2009), 'Australia's future tax system: the retirement income system. Report on strategic issues', Australian Government Department of Treasury, Canberra: Commonwealth Government, http://www.futuretax.gov.au.
Hogan, L. and B. Goldsworthy (2010), 'International mineral taxation: experience and issues', in P. Daniel, M. Keen and C. McPherson (eds), *The Taxation of Petroleum and Minerals: Principles, Problems and Practice*, New York: Routledge, pp. 122–62.
IFC (International Finance Corporation) (2001), 'Education sector strategy', Health and Education Department, International Finance Corporation, World Bank Group, http://www.ifc.org/ifcext/che.nsf/Content/StrategyE.
IIEP (International Institute for Education Planning) (2007), 'Higher education and development', *Newsletter*, **25** (1), www.iiep.unesco.org/fileadmin/user_upload/pdf/jane07.pdf.
IMF (International Monetary Fund) (2002), 'Fiscal dimensions of sustainable development', Pamphlet series No. 54, http://www.imf.org/external/pubs/ft/pam/pam54/contents.htm.
IMF (International Monetary Fund) (2011), 'World economic outlook report: tensions from the two-speed recovery', Washington, DC: International Monetary Fund, http://www.imf.org/external/pubs/ft/weo/2011/01/.
Jensen, S., B. Raffelhuschen and W. Leibfritz (1999), 'Public debt, welfare reforms and intergenerational distribution of tax burdens in Denmark', in A. Auerbach, L. Kotlikoff and W. Leibfritz (eds), *Generational Accounting Around the*

World, Chicago: University of Chicago Press, pp. 219–37, National Bureau of Economic Research, http://www.nber.org/chapters/c6692.

Kenyon, D. (2007), 'The property-tax-school funding dilemma', Lincoln Institute of Land Policy Focus Report, http://www.lincolninst.edu/pubs/1308_The-Property-Tax-School-Funding-Dilemma.

Kim, S. (2005), *Optimal Fiscal Policies for Sustainable Development: Greening the Tax and Budget System*, UNESCAP-KAPFE-KIPF Joint Symposium, 8–9 November 2005.

Kotlikoff, L. and L. Summers (1987), 'Tax incidence', in A. Auerbach and M. Feldstein (eds), *Handbook of Public Economics*, Amsterdam: North Holland, pp. 1043–92.

KPMG (2011), 'Economic analysis of using the GST to reform taxes', http://www.cpaaustralia.com.au/cps/rde/xchg/cpa-site/hs.xsl/news-and-media-media-releases-gst-rate-increase.html.

Kumar, R. (1991), 'Taxation for a cyclical industry', *Resources Policy*, **17** (2), 133–48.

Kunzel, P., Y. Lu, I. Petrova and J. Pihlman (2011), '*Investment objectives of sovereign wealth funds: a shifting paradigm*', International Monetary Fund Working Paper No. 19, Washington, DC: International Monetary Fund.

Land, B. (2010), 'Resource rent taxes: a reappraisal', in P. Daniel, M. Keen and C. McPherson (eds), *The Taxation of Petroleum and Minerals: Principles, Problems and Practice*, New York: Routledge, pp. 241–62.

Lewin, K. and F. Caillods (2001), 'Financing secondary education in developing countries: strategies for sustainable growth', International Institute for Educational Planning, UNESCO, http:// www.unesco.org/iiep.

Liu, Z. (2010), 'Viewpoints: United Nations Sustainable Development Journal', *Natural Resources Forum*, **34**, 85–8.

McDonald, I. (2005), 'Intergenerational equity', *Agenda*, **12** (1), 3–18.

McKibbin, W. and A. Stoeckel (2011), 'Global fiscal consolidation', Lowy Institute for Public Policy Working Papers in International Economics No 1, http://www.lowyinstitute.org/Publication.asp?pid=1582.

McKinsey Global Institute (2010), 'Debt and deleveraging: the global credit bubble and its economic consequences', http://www.mckinsey.com/insights/mgi.aspx.

McPherson, C. (2010), 'State participation in the natural resource sectors: evolution, issues and outlook', in P. Daniel, M. Keen and C. McPherson (eds), *The Taxation of Petroleum and Minerals: Principles, Problems and Practice*, New York: Routledge, pp. 263–88.

Mikesell, R. (1989), 'Depletable resources, discounting and intergenerational equity', *Resources Policy*, **15** (4), 292–6.

Ministry of Education, Bangladesh (2010), 'National education policy 2010', Dhaka: Ministry of Education, Government of the People's Republic of Bangladesh, http://www.unescodhaka.org/publications/books/english-version-of-the-national-education-policy-2011.

Mirrlees, J. (2011), 'Professor Sir James Mirrlees on poverty and redistribution', Asian Development Bank, http://beta.adb.org/features/professor-sir-james-mirrlees-poverty-and-redistribution.

Musgrave, R. and P. Musgrave (1989), *Public Finance in Theory and Practice*, 5th edn, New Zealand: McGraw Hill.

Norwegian Ministry of Finance (2007), 'Background note: Norway's position in relation to the international debate on sovereign wealth funds', http://www.regjeringen.no/en/dep/fin/tema/statens_pensjonsfond/norways-position-in-the-debate-on-sovere.html?id=487466.

OECD (2010), 'What are the economic benefits of education?', http://www.oecd-ilibrary.org/education/education-at-a-glance-2010/what-are-the-economic-bene fits-of-education_eag-2010-11-en.

OECD (2011), 'World economic outlook: November 2011', http://www.oecd.org/document/18/0,3746,en_2649_33733_20347538_1_1_1_1,00.html.

OECD (2012), 'OECD's current tax agenda 2012', http://www.oecd.org/data oecd/38/17/1909369.pdf.

Parry, M., O.F. Canziani, J.P. Palutikof, P.J. van der Linden and C.E. Hanson (eds) (2007), *Contribution of Working Group II to the Fourth Assessment Report of the Intergovernmental Panel on Climate Change: Impacts, Adaptation and Vulnerability*, Intergovernmental Panel on Climate Change, Cambridge: Cambridge University Press.

Pressman, S. (2002), 'Fiscal policy and the equity-efficiency trade-off', USBIG Discussion Paper No. 21, February, http://www.usbig.net/papers/021-Pressman.doc.

Rawls, J. (1999), *A Theory of Justice*, Cambridge, MA: Harvard University Press.

Reiche, D. (2008), 'Sovereign wealth funds as a new instrument of climate protection policy?', Wuppertal Institute for Climate, Environment and Energy, http://ipac.kacst.edu.sa/eDoc/2010/190460_1.pdf.

RBA (Reserve Bank of Australia) (2011), 'Statement on monetary policy', http://www.rba.gov.au/publications/smp/.

Rydell, I. (2005), 'Equity, justice, intergenerational transfers and the ageing population', Institute for Futures Studies, http://ideas.repec.org/p/hhs/ifswps/2005_005.html.

Sandbu, M. (2006), 'Natural wealth accounts: a proposal for alleviating the natural resource curse', *World Development*, **34** (7), 1153–70.

Sarker, T.K. (2003), 'Incidence of income taxation in Bangladesh', Keio University, Japan, Masters' thesis.

Sarker, T. (2007), 'Who bears the burden of taxes in developing countries? A case of income taxation in Bangladesh', *Pakistan Economic and Social Review*, **44** (2), 181–200.

Sarker, T. (2011), 'Enabling post-GFC mineral tax policy reform through strategic stakeholder communication', *International Journal of Trade, Economics and Finance*, **2** (5), 443–9.

Sarker, T. (2011), 'Time for a property tax: how it may work', *The Financial Express*, 11 March, http://www.thefinancialexpress-bd.com/more.php?news_id=128729&date=2011-03-11.

Sarker, T. and E. Whalan (2010), 'Reform issues in mineral tax policy for economic development: the mineral resource rent tax in Australia', *Bulletin for International Taxation*, **65** (1), 1–25.

Sarker T., M. Azam and S. Hossain (2012), 'Climate change adaptation in the coastal region of Bangladesh: current strategies and governance challenges', Planet Under Pressure 2012, 25–29 March, Excel London, forthcoming.

Scholl, G., F., Rubik H., Kalimo K. Biedenkopf and O. Söebech (2010), 'Policies to promote sustainable consumption: innovative approaches in Europe', *Natural Resources Forum*, **34**, 39–50.

Sigma Project (2003), 'The Sigma guidelines: toolkit', Sustainability Accounting

Guide, United Kingdom: Sigma Project, http://projectsigma.co.uk/Toolkit/SIGMAGlobalReporting.pdf.
Slemrod, J. and J. Bakija (2008), *Taxing Ourselves*, 4th edn, Hong Kong: MIT Press.
Smith, A. (1776), *The Wealth of Nations*, cited in A.B. Atkinson and J.E. Stiglitz (1980), *Lectures on Public Economics*, New York: McGraw-Hill.
Soltau, F. (2006), 'Introduction: climate change and sustainable development – understanding the linkages', *Natural Resources Forum*, **30**, 253–5.
Southalan, J. (2011), 'What are the implications of human rights for minerals taxation?', *Resources Policy*, **36**, 214–26.
Stanton, E. and F. Ackerman (2009), 'Climate and development economics: balancing science, politics and equity', *Natural Resources Forum*, **33**, 262–73.
Sunley, E., Baunsgaard, T. and D. Simard (2002), 'Revenue from the oil and gas sector: issues and country experience', Washington, DC: International Monetary Fund, http://siteresources.worldbank.org/INTTPA/Resources/SunleyPaper.pdf.
Swan, W. (2010), 'Australia to 2050: future challenges: Intergenerational Report 2010', Canberra: Australian Government Department of Treasury, http://www.treasury.gov.au/igr/igr2010/report/pdf/IGR_2010.pdf.
Takayama, N. and Yukinobu Kitamura (1999), 'Lessons from generational accounting in Japan', *American Economic Review*, **2**, 171–5.
Tanzi, V. and H. Zee (2001), 'Tax policy for developing countries', International Monetary Fund Economic Issues Paper No 27, http://www.imf.org/external/pubs/ft/issues/issues27/index.htm.
Thompson, J. (2002), 'Intergenerational equity: issues of principle in the allocation of social resources between this generation and the next', Australian Parliamentary Research Paper No. 7, http://www.aph.gov.au/library/pubs/rp/2002-03/03RP07.htm.
Thornton, K. (2012), 'Riding the solar-coaster', ABC Environment, http://www.abc.net.au/environment/articles/2012/03/07/3447005.htm, accessed May 2012.
UNESCO (2008), 'National report: adult education and learning in China: development and present situation', Chinese National Commission for UNESCO, http://unesdoc.unesco.org/Ulis/cgi-bin/ulis.pl?catno=213585&set=4EB2C715_0_460&database=ged&gp=0&lin=1&ll=Use.
United Nations (2011), *Human Development Report 2011*, http://hdr.undp.org/en/.
Usui, N. (2011), 'Tax reforms toward fiscal consolidation: policy options for the government of the Philippines', Asian Development Bank, http://beta.adb.org/publications/tax-reforms-toward-fiscal-consolidation-policy-options-government-philippines.
Van der Plog, F. and A. Venables (2011), 'Harnessing windfall revenues: optimal policies for resource rich developing economies', *The Economic Journal*, **121** (3), 1–30.
Wang, W., X. Zheng and Z. Zhao (2011), 'Fiscal reform and public education spending: a quasi-natural experiment of fiscal decentralization in China', *Publius: Oxford Journals*, http://publius.oxfordjournals.org/content/early/2011/10/12/publius.pjr039.full.
WCED (1987), *Our Common Future*, Oxford: Oxford University Press.
Weinthal, E. and P. Luong (2001), 'Energy wealth and tax reform in Russia and Kazakhstan', *Resources Policy*, **27**, 215–23.
Weiss, M. (2008), 'Sovereign wealth funds: background and policy issues for congress', CRS Report for Congress, http://www.crs.gov.au.

World Bank (2011), 'Education and development', http://go.worldbank.org/F5K8Y429G0.

World Bank (2011), 'Poverty at a glance: Bangladesh', http://www.worldbank.org.bd/WBSITE/EXTERNAL/COUNTRIES/SOUTHASIAEXT/BANGLADESHEXTN/0,,menuPK:295785~pagePK:141132~piPK:141109~theSitePK:295760,00.html.

Xinhua (2010), 'Bangladesh desperate for foreign loans to meet budget deficit', 18 April, http://english.peopledaily.com.cn/90001/90778/90858/90863/6954688.html.

5. Indian agriculture in the era of global warming

Ranajit Chakrabarty and Smwarajit Lahiri Chakravarty

INTRODUCTION

The steady and continuous rise of the average temperature of the earth's surface, alterations in the radiation balance of the earth–atmosphere system and the subsequent consequences are usually referred to collectively as global warming. As per the Third Assessment Report (TAR) of the Intergovernmental Panel on Climate Change (IPCC 2001), the temperature of the earth's surface has increased by about 0.6°C, ±2°C in the twentieth century. By the twenty-first century, the earth's temperature is expected to increase by 1.6°C to 2.0°C.

The root cause of this abrupt change in the climate is the increase in the concentration of greenhouse gases (GHGs), mainly carbon dioxide and methane, in the upper atmosphere of the earth. The extra gases that build up in the earth's atmosphere prevent additional thermal radiation from leaving the earth and trap the excess heat of the sun. The total carbon dioxide emitted in the world is 27 billion tonnes (Manne and Richels 1993). Globally, agricultural methane and nitrous oxide emissions have increased by 17 per cent from 1990 to 2005, and are projected to increase by another 35–60 per cent by 2030, partly driven by the intensive use of nitrogen fertilizer and increased livestock production. This increase in the GHGs is due to human activities like burning of fossil fuels, industrial processes, farming and deforestation (Oak Ridge National Laboratory 1993).

In 1994, according to the Government of India, the total emission of all GHGs from India, expressed in CO_2 equivalent terms, was 1228 million metric tonnes (mmt), of which carbon dioxide accounted for 65 per cent. Thus India was the fifth largest emitter of CO_2. Approximately 85 per cent of CO_2 emissions came from the energy sector, 13 per cent from industrial processes, and the remaining 2 per cent from land use, including changes therein. Methane accounted for 31 per cent of total GHGs (Murthy et al. 2000). Its major source was rice cultivation, enteric fermentation by cattle, energy sources such as biomass burning, coal mining and handling, flaring of natural gas, and waste-disposal activities. Nitrous oxide formed 4 per cent of total GHGs, originating in agriculture, fuel combustion, industrial processes, and waste and biomass burning (Mitra 1992).

The models derived from climate change data show the average temperature of India will rise by 2 to 5° Celsius in the next 50 to 100 years. This abrupt change in the climate will lead to intense rainfall, an increase in the rate of the melting of glaciers, a rise in sea level and an increase in extreme weather phenomena, which is likely to result in frequent floods and severe drought in some parts of the country. The change in the weather may have an adverse impact on the water balance of the country. The salinity of sea water is expected to change due to the increase in the sea level.

There is a basic difference between land and ocean temperature increase. Since 1980, the land temperature has increased at double the rate of the temperature increase of seas and oceans. This can be explained by the fact that the heat capacity of the ocean bodies is an effective instrument while at the same time the element of evaporation cannot be ignored. This idea would suggest for example that the southern hemisphere heats up more slowly than the northern hemisphere because the former has more prominent and larger water bodies than the latter.

Surface warming, along with altered water balance, is expected to have a severe impact on the productivity of crops. It is expected that some crops may benefit due to the increase in the carbon dioxide in the atmosphere. The overall impact of global warming on agriculture is negative, as the productivity of crops is likely to decline in near future, which may threaten the country's food security. A drastic change is also inevitable in the forest ecosystem, which is expected to experience the decline of forest area and types. This change may result in the potential loss of forest biodiversity.

In the last two decades, the global warming issue has moved from scientific circles to policy circles (Mongia 2007). Policymakers across the globe are seriously trying to derive the best set of responses to deal with the difficult situation that has evolved, which could ultimately lead to the extinction of humans. In the fourth assessment report of the IPCC (2007) it was clearly stated that warming of the climate system is unequivocal at present, due to increases in global average air and ocean temperatures, widespread melting of snow and ice and the rise in global sea level.

The international apex body on climate change is the United Nations Framework Convention on Climate Change (UNFCCC) which stresses adaptation strategies with respect to climate change only. As previously mentioned, it is the emission of GHGs that leads to climate change and hence the need for adaptation becomes a must. It is also noted that adaptation becomes vital as mitigation of GHGs might not hinder climatic changes (Stern 2006).

Hence, it follows as a natural consequence that for mitigation and adaptation policy implementations, one needs to study the possible effects on the various important sectors. Such methodologies, with respect to mitigation, would give us the needed logic for the de-carbonizing of energy systems. Along with this, it is vital to recognize the important regions and sectors so that the knowledge of adaptation may be used.

LITERATURE REVIEW

The idea of global warming is not new but nearly 200 years old. The earliest concept was proposed by Fourier (1827), who thought that the atmosphere was a hot house that allowed the 'light rays' of the sun but did not allow the 'dark rays' of the earth to leave. John Tyndall (1861) suggested that carbon dioxide acts as a good absorber of water vapour. Arrhenius (1896) calculated that 8°C rise in global temperature would be expected if the global carbon dioxide doubled. Chamberlin (1897) developed the carbon dioxide theory for calculating climatic changes. Revelle and Suess (1957) gave a solid foundation regarding the basis of global warming following the increase in carbon dioxide. The Conservation Foundation (1963, noted in Ausubel 1983) calculated that a 3.8°C temperature rise could follow the doubling of atmospheric carbon dioxide. The US President's Scientific Advisory Committee (1965) discussed global warming in its environmental pollution report, with apprehension.

The Study of Critical Environmental Problems (SCEP) in 1970 reviewed the global warming problem (Viskanta 1976). Coope et al. (1971) stated that while global warming was in question, cooling was also held as a strong possibility. The Villach Conference in Austria (1985) simply stressed the urgent need to tackle the global warming problem (United Nations Environment Programme 2002). Rowlands (1992) stated that the strategy to reduce GHG emissions was taken up seriously from the mid-1980s.

The establishment of the IPCC in 1988 followed as a natural consequence after the Toronto Conference. The IPCC (1996) report provided methods to depict trends in temperature and precipitation. Two time series data sets are usually used. The land surface air temperature was developed by James (1994, pp. 266–268) while that of the water bodies and oceans was developed by Folland et al. (1984). These data sets are used to calculate the temperature changes that were reported in the second

Assessment Report (SAR) of the IPCC in 1996. Temperature records all around the globe from 1861 to 1994 are provided by the IPCC for a broad number of contiguous regions. According to SAR, the average global surface temperature has risen by 0.5°C since 2001. This may be due to other factors as well and may not be wholly attributed to the carbon dioxide increase. Scholars have found positive correlation between carbon dioxide concentration and temperature anomalies, but this could be spurious (devoid of a cause–effect relation). Friss and Lassen (1991) used interglacial periods and changes in intensity of solar activity to explain the global warming trends. Laut and Gundermann (1998) and Lockwood et al. (1999) stated that anthropogenic climatic change is dominating and that global warming cannot be explained only by solar activity even though it is prominent.

The Third Assessment Report (TAR) (IPCC 2001) stated that over the twentieth century, the global average surface temperature increased by 0.6°C. The maximum heating took place from 1976 to 2000. Since the late 1950s, the overall rise in global temperature was about 0.1°C per decade. For India, for the period 1901–2000, a very low decadal rise of 0.2°C was recorded. Thus, there was a continuous upward trend.

AGRICULTURE

Agronomic research suggests that a rise in temperatures will have a harmful impact on many crops. The crops growing in water stressed or heat stressed areas, or in areas with a combination of the two, are vulnerable to various diseases. The rise in temperature of the atmosphere is likely to affect the production of rice (Lal et al. 1998), thereby affecting the major rice growing countries of the Asian sub-continent.

If we look at the projections up to the year 2100, then for the Indian case there is an overall increase in temperature by 2–4°C with an increase in precipitation. Mall et al. (2006) analysed and showed that there would be a drop in the yields of cereals like rice and wheat under climate change conditions.

Reilly (1996) and Reilly and Schimmelpfennig (1999) specified that the change in climate would have an impact on crop productivity and yields in tropical countries, but that countries which are at higher latitude, such as European countries, are expected to benefit due to the rise in atmospheric temperature. The presence of moisture in the soil during the growing season of the crop is a very important factor for better productivity. An increase in global temperature may increase the tendency of frequent droughts and heavy rains, thereby damaging the crops. Irrigation, an important part of the agriculture, will also be affected by water scarcity caused by global warming. The hotter summer and spring will increase the rate of evaporation, thereby causing drier soil.

Another expected effect of global warming is increased susceptibility of crops to pests, which would result in the usage of more pesticides. The productivity of crops will also be affected by increased heat stress, frequent flooding and increased salinity of the soil due to the rise in sea level.

It has been predicted (University of Reading 2000) that in the next 50 years the impact of global warming on agriculture will be that:

1. The carbon dioxide in the atmosphere will increase from 360 PPM to 450–600 PPM, which is expected to be good for crops as the rate of photosynthesis will increase and the use of water will be reduced.
2. The sea level will rise by 10–15 cm, and hence land will reduce, coastal erosion will increase, there will be frequent floods and the salinity of the groundwater will increase.
3. The atmospheric temperature is likely to rise by 1–2°C, which will result in frequent heat waves. As a consequence there will be a shorter and earlier growing season. There will also be a risk of heat stress, and transpiration will increase.
4. There will be frequent storms and rainfall will increase, which will lead to soil erosion, and infiltration in the rainfall will increase.

IMPACT OF GLOBAL WARMING ON INDIAN AGRICULTURE

India is a large country with 15 agro-climatic conditions and different kinds of soil, which is favourable for the cultivation of varieties of crops. The country follows different methods of farming in different climatic zones. Agriculture has been the primary source of income for a major section of the people of India since time immemorial. At present about 65–70 per cent of the country's workforce are employed in this sector (Misra and Puri 2010). Agriculture and animal rearing generate substantial support for the country's economic growth as well as for the social transformation of the country (Tables 5.1 and 5.2).

Dev (2007) observed that the growth of the agriculture sector is remarkable in the post independence period compared to that of the pre independence period. From 1949/50 to 1999/2000, the growth of production was 2.7 per cent per annum, which is much higher than the growth rate of 0.4 per cent in the first half of the twentieth century. The Green Revolution made the country secure in food grains.

Table 5.1 Share of agriculture GDP in total GDP: all India (at 1993–94 prices)

Year	Agriculture, forestry, and fishing	Agriculture*
1950–51	57.7	50.2
1960–61	53.0	47.3
1980–81	39.7	35.8
2002–03	22.0	20.0
2010–11	16.6	14.6

Note: * Refers to crop production and livestock.

Source: National Accounts Statistics of India: 1950–51 to 2002–03 (2004); Economic and Political Weekly Research and Foundation, December 2004, Mumbai.

Table 5.2 Share of agriculture in total employment: all India

Year	Agriculture, forestry, and fishing	Crop production and plantations
1961	75.9	73.5
1999–2000	59.9	54.3
2000–10	53.6	50.2

Source: Sundaram (2001); Economic and Political Weekly, 11 August: 3039–49.

The growth rate of GDP from the agricultural and allied sectors was highest in 1980; it was more than 3 per cent (Table 5.3). But it declined to 2.76 per cent per annum in the post reform period. The decline in the growth rate of agriculture from the first half of the 1990s onwards has become a matter of concern. During 1997/98 to 2004/05, agricultural growth was only 1.6 per cent per annum.

Extensive cultivation characterized Indian agriculture during the pre-1965 era, and intensive cultivation in the post-Green Revolution period. During the pre-1965 era the country featured wide-ranging agricultural productions. During the post-Green Revolution period, the use of modern inputs like man-made fertilizers, insecticides, pesticides etc. increased many fold, which resulted in huge production. During 1950/51 to 2003/04, the percentage of net irrigated area to net cultivated area increased from 17 per cent to 41 per cent. The consumption of fertilizer showed a significant increase from 1 kg/hectare to 90 kg/hectare, which increased the production while the quality of the produce fell. The per-

Table 5.3 Growth rates in agriculture GDP: all India

Period	Growth rate (% per annum)
1950–55 to 1964–65	2.51
1967–68 to 1980–81	2.20
1980–81 to 1990–91	3.07
1992–93 to 2001–02	2.76
1992–93 to 1996–97	3.85
1997–98 to 2004–05	1.60
2004–05 to 2010–11	1.41

Notes: GDP is in 1980–81 constant prices from 1950–51 to 1980–81; in 1993–94 constant prices for the period 1980–81 to 2001–05.

Source: National Accounts Statistics (various years); Central Statistical Organization, Government of India.

centage of area of high yielding varieties of cereals rose from 15 per cent in 1970/71 to 75 per cent in 1990/91. The consumption of electricity in agriculture has also risen from 4 per cent in 1950/51 to 30 per cent in the present time.

The increased rate of crop production has also led to the increase in agricultural area. The post-Green Revolution period does not show any significant increase in the agricultural growth rate. In the 1980s the yield growth rate was highest, at 2.56 per cent per annum, and declined drastically to 1.31 per cent in the 1990s. The production as well as the yield of the food grains and non-food grains was highest in 1980, and again shows a significant fall in the 1990s.

Output of all crops increased at a compound growth rate of 3.15 per cent per annum during 1949/50 to 1964/65, significantly led by area expansion (Table 5.4). There has not been any significant increase in agricultural growth in the post-Green Revolution period. However, the sources of growth have changed from area expansion in the pre-Green Revolution period to yield growth in the later periods.

The aggregate output growth at national level hides a great deal of variation in the performance of different crops and regions. While wheat has recorded high and accelerated rates of yield increases, others like coarse cereal and pulses have not. Food grains and non-food grains recorded the highest growth rates in output as well as yields in the 1980s. In the 1990s, the growth rate decelerated for both food grains and non-food grains. Increasing commercialization in agriculture started in a significant way in the 1980s.

Table 5.4 Compound growth rates of area, output and yield: all India

Period	All crops			Food grains			Non-food grains		
	Area	Output	Yield	Area	Output	Yield	Area	Output	Yield
1949/50 to 1964/65	1.58	3.15	1.21	1.35	2.82	1.36	2.44	3.74	0.89
1967–68 to 1980–81	0.50	2.19	1.28	0.38	2.15	1.33	0.94	2.26	1.19
1980–81 to 1989–90	0.10	3.19	2.56	−0.23	2.85	2.74	1.12	3.77	2.31
1990–91 to 1999–2000	0.25	2.28	1.31	−0.17	1.94	1.52	1.37	2.78	1.04
1999–2000 to 2000–10	0.22	2.08	1.11	−0.16	1.84	1.42	1.33	2.68	1.01

Source: Ministry of Agriculture, Government of India (various years).

CLIMATE CHANGE AND AGRICULTURE

In general the economic impacts of climatic change on agriculture have been studied in great detail. The existing literature speaks about two broad approaches:

1. The crop modelling or production function approach stresses the structural modelling of crop and farmer response along with the decisions of farmers.
2. The spatial analogue or the Ricardian approach analyses the differences in agricultural production and climate among different regions.

The basic idea of the first approach is that physical impacts are exogenously fitted into an economic model as Hicks-neutral technical changes. In India, as estimated by Kumar and Parikh (2001), if CO_2 concentration levels are doubled in the latter half of twenty-first century, GDP would decline by 1.4–3.0 per cent under climatic changes. Along with this, it was shown that an increased proportion of the Indian population would be in the lower income groups. The major difficulty in this regard is the treatment of adaptation. Under the different physical agricultural impacts each adaptation strategy changes, and so, for the policymaker, the choice set of the selection of a particular strategy becomes limited.

Mendelsohn et al. (1994) used cross-sectional US data and with econometric models, under the Ricardian approach, tried to link land values to climate. The approach is better as it can include all adaptation policies, given that the analysis is based on the observed behaviour of the farmer.

The first approach does not consider the response of farmers as a key element. It is considered vital by the Ricardian approach, where the farmers have perfect foresight as they can identify the changes in climate instantly.

Objectives of the study

In this exercise, we try to determine and estimate effects of adverse climatic changes on agriculture. Hence we ask:

1. What is the effect on the output of rice and wheat due to climatic changes?
2. If both temperature and precipitation rise, what would be the dynamics of such changes?

MODEL SPECIFICATION

In this exercise, we try to introduce the Ricardian model (Kumar and Parikh 1998). In a less developed country like India, the non-availability of data is a major obstacle. As a result, we have to use a variant of the Ricardian model which is based on the availability of the data. We have used net productivity (NP) as a social indicator to estimate the welfare changes. The effect and changes in the environment are neatly captured through NP. So, the functional relation of the Ricardian model is specified as follows:

$$NP = f\,(T_j, R_j\ h, pdn, a, cr) \qquad (5.1)$$

where:

T_j ➔ represents the temperature in °C
R_j ➔ represents the rainfall in mm
j ➔ stands for the season

It must be understood that based on the goodness of fit, we have introduced a quadratic relation that would serve as the regression equation.

Equation (5.1) also has the control variables which are:

h ➔ height of that region given in m,
pdn ➔ population density,

a → net area under cultivation,
cr → credit available, which captures the purchasing power to buy tractors, cows, bullocks, fertilizers, pumps, etc. needed for agricultural output.

The data source is rich but unfortunately, there are some missing data points. These empty spaces have been filled up by using interpolation or else the missing value analysis might give us wrong results. The period of study considered is from 1950 to 2003. For the purpose of comprehensiveness, we have covered all the four zones of India, selecting the states as follows:

North – Uttar Pradesh (undivided), Punjab, Madhya Pradesh
South – Andhra Pradesh, Karnataka, Tamilnadu
West – Gujarat, Maharashtra, Rajasthan;
East – West Bengal, Bihar (undivided) and Orissa

Thus, we have selected three major states in each zone, and hence we are considering a total of 12 states. The remaining states are not so significant and they suffer from extreme conditions. For example, states in the north east have relatively smaller contributions whereas states like Jammu and Kashmir, Himachal Pradesh suffer from extreme climatic conditions. Hence, the states that we have selected represent a balanced approach.

The study basically covers the net area under cultivation and we have chosen only the major crops – rice and wheat. While rice is favoured in both east and south India, wheat is the favourite of west and north India. The credit available to the farmers includes the costs of farm mechanization. This in turn incorporates labour cost also as it is difficult to determine the time labour cost, especially when the farmer supplies his own labour. It is difficult to identify the prices of the key agricultural inputs and so these are considered as control variables.

The data on climate, that is, temperature, is available from 1901 to 2003, but the exercise considers the time period from 1950 to 2003; that is, a 54 year analysis is carried out. The data is available for all-India levels as well as for north, south, west and, east regions. So, the regional data is used in this analysis. The data source is the Indian Institute of Tropical Meteorology (IITM), Pune. Following the Indian meteorological departments, we have subdivided the entire year into four parts.

1. 1st part: January, February, March
2. 2nd part: April, May, June

3. 3rd part: July, August, September
4. 4th part: October, November, December

It is with respect to these subdivisions that the climatic variables are introduced and explained where rainfall and temperature are the other variables under study.

For the purpose of our study, the data set has been divided into four parts.

1. 1950–1965 ➔ the stage after independence with slow reforms in agriculture.
2. 1966–1980 ➔ the period of Green Revolution with major reforms in agriculture
3. 1981–1991 ➔ the stage where economic liberalization was initiated with reforms
4. 1992–2003 ➔ the post liberalization period

In this exercise, we take up the effects of the climate/temperature changes over the two specific agricultural products, that is, rice and wheat.

HAUSMAN TEST[1]

If in a model and data we have fixed effects estimation as appropriate, the Hausman test tries to find out whether random effects estimation would be almost as good as the fixed effects. In fixed-effects estimation, the Hausman test considers the null hypothesis that random effects would be consistent and efficient against the alternative that random effects would be inconsistent. In most cases, the fixed effects are consistent. The test statistic is distributed chi-square. If the Hausman test statistic is large, one must use the fixed effects model, but if the statistic is small, one may use the random effects model. It is noted:

1. If we want to control for omitted variables that differ between cases but are constant over time, we would use the fixed effects regression. This allows us to incorporate the changes in the variables over time to estimate the effects of the independent variables on the dependent variable.
2. If we want to control for omitted variables that change over time but are constant between cases, we would use the regression model with between effects. This helps us to estimate the effect of the omitted independent variables on the dependent variable by using the variation between cases.

3. If we feel that some omitted variables may be constant over time but vary between cases, and others may be fixed between cases but vary over time, then we may include both types by using random effects.

In choosing between fixed and random effects, we would have to run the Hausman test. Fixed effects are always consistent but may not be the most efficient model to run. Random effects will give us an efficient estimator, so we should run the random effects. Against a less efficient but consistent model, the Hausman test verifies for a more efficient model to ensure that the more efficient model also gives consistent results. We run the Hausman test comparing fixed with random effects and then carry out the comparison. We formulate the null hypothesis that the coefficients estimated by the efficient random effects estimator are the same as the ones estimated by the consistent fixed effects estimator. If they give insignificant P-value then random effects may be used. We should use fixed effects if we get a significant P-value.

To carry out the exercise, we now depend on the Hausman Test. The basic idea of this approach is to compare two sets of estimates whereby one is consistent only under the null hypothesis whereas the other is consistent under both the null and alternative hypothesis. A large difference between the two sets of estimates is taken as evidence in favour of the alternative hypothesis. The original exercise considered the proposal of a test statistic for endogeneity based upon a direct comparison of coefficient values. The test further modified by Davidson and Mackinnon is used here.

Let us assume that we are interested in the fact the y is endogenously determined with x where x is derived through some prespecified functional relation. The ordinary least squares (OLS) estimates would be biased and inconsistent if endogeneity is present. Thus, in order to test this hypothesis, we need to find a set of instrumental variables that are correlated with the variable y. Let us consider that z_1 and z_2 are two such instruments.

We can now carry out by artificial regression the two OLS regressions. In the first regression equation we regress the suspect variable y on all the exogenous variables and instruments and retrieve the residuals

$$Y = a + bx + z_1 + z_2 + u_1$$

We then, in the second regression equation, re-estimate x by including the residuals from the first regression as additional regressors.

$$X = c + dy + \text{residual y} + u_2$$

If the OLS estimates turn out to be consistent, then the coefficient on the first stage residuals should not be significantly different from zero.

When we apply this Hausman test in our analysis, we are actually testing for both fixed and random effects for each of the four time periods. In each case we find that the model consisting of random effects generates estimates that are biased which implies that the null hypothesis is rejected. That is, the natural consequence that follows is that we have to favour the fixed effects estimators.

The analysis of the 54 year period is expected to give us sufficient good interpretations of the changing nature of climate/temperature on the agricultural output of two crops – rice and wheat. Also, each state differs from another with respect to geographical location, geographical size, etc. So, the production of rice and wheat is weighted by the total area under these two crops.

ANALYSIS

The basic idea of our analysis is to check whether the changing nature of climate sensitivity is in any way affecting agricultural output or not. The period of analysis is 1950–2003, divided into four sub periods as already mentioned. Thus, we are trying to estimate Equation 5.1.

The estimation procedure is based on the fact that we are using fixed effects criteria against the random effects as clarified by the Hausman test. The estimation procedure isolates the climate coefficients for the four sub periods whereby the pooled data is used for the period 1950–2003.

We find that all the estimates are derived on the concept of fixed effects specification ($R^2 = 0.813$).

Table 5.5 gives us the estimates of the climate coefficients. The dependent variable in each case is the net productivity which is expressed at constant prices, considering 1993/94 as the base year. The analysis generates some interesting results.

1. In all the four periods we find that the control variables have the expected sign.
2. Ignoring the few exceptions, the climate coefficients are all significant.
3. Regarding the climate coefficients, the F tests rejected the null-hypothesis that climate coefficients do not affect agriculture.
4. It is obvious that the climate coefficients are affecting agriculture in an adverse direction.

Table 5.5 Estimates of the coefficients under Hausman test

	1950–65		1966–80		1981–91		1922–2003	
	Coefficient	P value	Coefficient	P value	Coefficient	P value	Coefficient	P value
Jan T	−337.8	0.00	−238.9	0.00	−403.2	0.01	−483.2	0.02
Apr T	−123.9	0.20	−679.3	0.01	−793.2	0.00	−642.3	0.01
Jul T	−613.8	0.00	−683.9	0.00	−687.2	0.02	−587.9	0.00
Oct T	486.7	0.00	776.3	0.00	987.2	0.01	876.3	0.03
Jan R	23.9	0.21	32.9	0.00	113.2	0.00	48.9	0.01
Apr R	−7.1	0.04	−11.3	0.01	−14.3	0.02	−7.2	0.02
Jul R	−3.2	0.76	−15.2	0.00	1.2	0.00	2.1	0.00
Oct R	28.9	0.00	−38.9	0.02	19.3	0.01	−6.2	0.01
Jan T^2	−1.2	0.81	−53.9	0.03	23.3	0.00	26.1	0.01
Apr T^2	−18.6	0.61	155.7	0.02	52.7	0.00	13.3	0.02
Jul T^2	−176.4	0.00	−93.9	0.00	−423.1	0.03	−112.1	0.03
Oct T^2	−187.3	0.00	−223.2	0.00	−326.9	0.02	−416.9	0.04
Jan R^2	−0.9	0.05	−82.1	0.01	−4.1	0.00	−3.1	0.01
Apr R^2	0.3	0.00	0.4	0.02	−0.8	0.01	0.7	0.00
Jul R^2	0.1	0.02	0.2	0.00	0.5	0.02	0.4	0.01
Oct R^2	−0.8	0.07	0.4	0.01	−0.2	0.00	0.1	0.00
Control variables								
H	14.6	0.11	18.2	0.12	19.8	0.11	18.6	0.00
pdn	93.8	0.31	97.3	0.01	102.3	0.14	107.3	0.01
A	142.3	0.12	168.9	0.13	155.3	0.13	157.3	0.13
cr	196.8	0.01	208.3	0.04	204.2	0.02	202.8	0.40

Table 5.6 Dynamic climatic changes on agriculture

Situation temperature/ rainfall	1950–65 % change NP wrt 1991/92	1966–80% change NP wrt 1991/92	1980–91 % change NP wrt 1991/92	1992–2003 % change NP wrt 1991/92
+ 1.5°C/5%	−3.1	−2.2	−7.3	−11.3
+ 3.0°C/10%	−12.8	−10.8	−15.8	−19.3

In this regard, the most vital concept is the net productivity which is derived as the difference in the year to year growth rates. It must be understood that in this exercise we have tried to link up the net productivity and climatic changes.

All agricultural output is not marketed. We consider the net agricultural output of rice and wheat that is either marketed or consumed which when divided by the total land area gives us the output per hectare or the net productivity (NP). The climate change effects are estimated by the changes in the NP which is given as quintals per hectare. It may be noted that as India is a vast country, the impact of climatic change is different over the different regions.

We then find out the impacts by considering the four zones separately. These are then added and averaged to give us some idea of national level impacts. The value derived may not reflect the all India figures but still it would be a good estimate of the same.

Thus, in this analysis, the effects of climatic changes are measured through the changes in net productivity. The average effects for all the years are mentioned in detail in Table 5.6. We select the output of 1991/92 at 1993/94 prices as the base years. Thus, the NP values are expressed at constant prices.

Liberalization programmes were introduced in 1991, but left the agricultural sector unchanged. So, taking 1991/92 as the base year would not affect the basic result of the exercise. Thus, with respect to this base year the impacts are calculated and then for each of the four periods, these average values are represented. Without a comparison with a particular benchmark year, the estimates and the results would have no meaning. The adverse effects are based on the changes in NP (of 1991/92) if future climate changes are assumed. In a dynamic scenario, we see that over time these effects are increasing which suggests increasing negative climatic change effects on agricultural output, especially with respect to rice and wheat. This pattern is uniform even though we accept that India has made huge strides especially in the agricultural sector.

If we look at the data more carefully, we find that the modulus (absolute) value of the effects is increasing. But in the period of the Green Revolution, these effects are subsequently low which may be due to better use of fertilizers, high yielding variety (HYV) seeds and implementation of better techniques of production. But in the first period and also in the third period we find the impacts are high, which may indicate that the vulnerability and weakness of Indian agriculture is growing over time. In the last period we find that the effects are much higher due to the neglect of agriculture in the post liberalization period.

SPATIAL AUTOCORRELATION

To get a very good estimate of the climate sensitivity of agriculture, it is necessary to account for spatial autocorrelation so as to improve upon the efficiency of the results. We can consider the spatial error model as given by equation 5.2 and the spatial lag model as given by equation 5.3. For sake of simplicity, we have considered a linear model

$$y_t = \alpha + \beta x_t + \rho W \eta_t + \varepsilon_t \tag{5.2}$$

where $\eta_t = \rho W \eta_t + \varepsilon_t$

$$y_t = Y + \rho W \eta_t + \beta x_t + \varepsilon_t \tag{5.3}$$

where

y_t → dependent variable
x_t → independent variable
β → slope of equation (5.2) or regression coefficient
η_t → correlated error terms
ρ → spatial autoregressive parameter
W → weight matrix
ε → error terms

It is noted that the independent variables include both the climate and the other control variables. It is thereby necessary to define the weights. The weights of each explanatory variable are in proportion to their representation in the model such that the sum of the weights is equal to unity.

We can now study the effects on climate sensitivity due to spatial autocorrelation. In this regard, we consider the period 1950–2003.

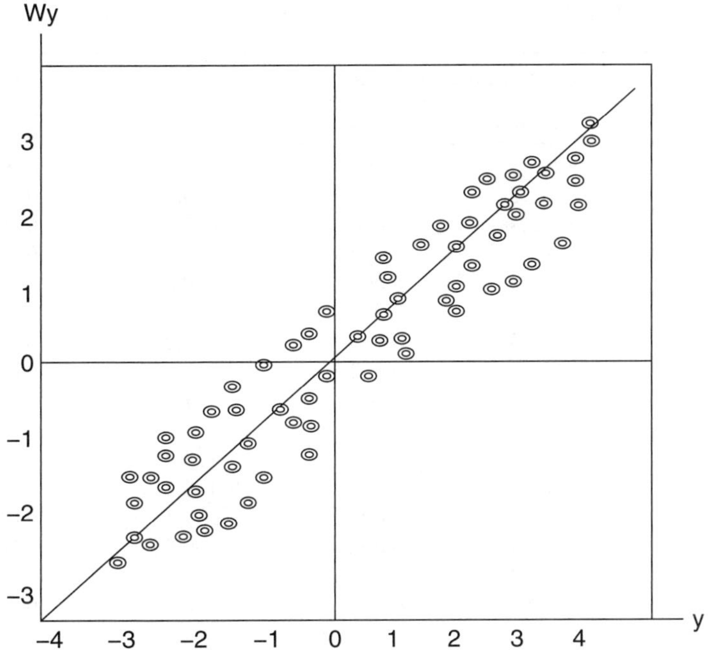

Figure 5.1 Moran's I = 0.468

In the plot the data of the NP values which is nothing but the dependent variable y, against Wy, we find that the scatter points are clustered in the north east and south west areas of the diagram/quadrants (Figure 5.1). This clearly specifies that there is positive spatial autocorrelation. The equations as specified by (5.2) and (5.3) are now estimated for the time period 1950–2003 as mentioned already. The results are presented in Table 5.7.

The estimates are generated by considering the fixed effects specification whereby the weights are calculated by considering the total area under cultivation for rice and wheat in this exercise. We find that there are two types of climate coefficients: one accounts for the autocorrelation while the other set is devoid of this autocorrelation. We find that the latter is higher in numerical terms than the former, in most cases. The analysis clearly tells us that there is sufficient presence of climatic sensitivity on Indian agriculture.

This is mainly due to the Indian farmer's limited knowledge along with limited or no education and also poor awareness of the environment. Thus, a changeover from spring to summer or from summer to the rainy

Table 5.7 Climate sensitivity affected by spatial autocorrelation

Variable	Without autocorrelation		With autocorrelation (lag model)	
	Coefficient	P value	Coefficient	P value
Jan T	−233.3	0.001	−198.7	0.000
Apr T	−437.5	0.000	−383.9	0.000
Jul T	−623.8	0.000	−462.3	0.000
Oct T	1087.2	0.002	876.9	0.001
Jan R	27.5	0.002	9.2	0.003
Apr R	−13.2	0.000	−10.8	0.002
Jul R	−1.3	0.000	−0.7	0.002
Oct R	21.4	0.000	18.8	0.001
Jan T^2	−33.8	0.001	−21.3	0.003
Apr T^2	98.7	0.000	87.4	0.000
Jul T^2	−76.9	0.002	−23.4	0.005
Oct T^2	−189.8	0.000	−148.7	0.000
Jan R^2	−1.3	0.000	−1.1	0.000
Apr R^2	0.9	0.000	0.8	0.000
Jul R^2	0.2	0.001	0.2	0.000
Oct R^2	0.3	0.113	0.1	0.000
Control variables				
h	246.3	0.000	163.9	0.000
Pdn	187.9	0.001	146.7	0.000
a	932.6	0.001	732.9	0.001
cr	443.7	0.002	287.0	0.000
Adjusted R^2		0.81		0.81

seasons is perceived by him as a major climate change. The traditional Indian farmer is still dependent on the monsoons. So, if the monsoons are late, he is worried and the agricultural output is affected. For him, the climatic concerns are floods, droughts, and so on that are directly and immediately linked with output.

If there are indeed major climatic changes, the farmers' scope for adaptation is limited. Moreover, there is a lack of flow of information. The government has introduced various channels for information flow, such as television, radio and the news media. Unfortunately, most Indian farmers do not have access to these and thereby the information flow, however limited in scope, becomes more or less useless. It has been observed that poor farmers depend on the big, rich and successful farmers for information and advice. It is these farmers that have access to the information provided by the government. But often it seems that these big, rich farmers

Table 5.8 Impacts of temperature and precipitation

Temp/rainfall	Without autocorrelation		With autocorrelation (lag model)	
	Impacts	% change NP	Impacts	% change NP
1.5°C/5%	−70.2	−8.76	12.2	1.38
3.0°C/10%	−95.7	−12.3	8.31	0.76

operate as 'gatekeepers'. So the flow of information from them to the small farmers is restricted.

There is another major problem in this regard. The information needs are not the same across farmers and depend on the types of crops cultivated and also on the farmer's scope and scale of operation. Thus, the information that is flowing in must pass through some proper and appropriate filters or else the process of reaching out to the farmers becomes quite ineffective.

In the period after the mid 1960s, the Indian farmer had the option of using better quality seeds and fertilizers. But the traditional Indian farmer is conservative and this complicates the situation. This can be analysed from Table 5.8.

We can now consider the effects with or without autocorrelation. We consider two hypothetical situations:

1. In the first case, we consider a 1.5°C increase in temperature and a 5 per cent increase in precipitation.
2. In the second case, we consider a 3.0°C increase in temperature and a 10 per cent increase in precipitation.

The impacts clearly speak about the effects that are generated and are observed as the percentage change in NP. If we consider the model without autocorrelation, we find that the values are negative but if autocorrelation is considered, the values are positive. We find that the lag model gives us positive changes though in numerical terms these are not very strong.

RESPONSE OF SENSITIVITY ANALYSIS

The inclusion of spatial autocorrelation effects has led to positive impacts as it has led to lowering of climatic effects. It is important to look deeply into the reasons for such impacts because this would in the long run help us to set up the proper policy framework. There is an explanation that communication across farmers (inter farmer) could be one of the major reasons in this regard. Thus, it is crucial to know about farmers' ideas about climate change and also the various routes by which they get such information.

It has been observed that most farmers are not aware of the idea of climatic changes (Parikh and Parikh 1998). The information flow on the use of fertilizers is also limited. So, the farmers again have to depend on the rich, big farmers for information regarding the type of fertilizer and pesticide that must be used. Sometimes, they collect this information from the dealers and sellers.

Most of the farmers in India are poor and therefore they do not have the necessary funds to buy these fertilizers, along with tractors and other machine equipment. The idea of co-operatives came up in the 1970s, whereby the pooled resources of various farmers were used. But in India the concept of co-operatives did not pick up and was not hugely successful.

At the same time, one should focus on the agricultural labourers also. In less developed countries like India, there is no proper mechanism to train them in agricultural practices and so it is the unskilled people that are involved in such processes.

An additional problem that emerges is that of uneconomic holdings. Many agricultural plots are so small that proper procedures of mechanization cannot be applied, and so the traditional methods of farming are still carried out. These types of uneconomic holdings actually lower the agricultural productivity, and the traditional methods lower it even further.

The corporate sector to date has not shown any interest in involving itself with the agricultural sector. They might take up these problems and tackle them by providing extension services, given that the government cannot manage it all alone. Also, their involvement would provide this sector with the required impetus, given that most industries have been set up in areas which were previously agricultural plots. The private sector, through R&D efforts, could help the agricultural sector. Also, it can transfer technological know-how to farmers which in turn becomes helpful in the following ways:

1. It lowers the cost
2. Output and NP rises
3. Profit of the agricultural sector improves
4. Consumers get the products at cheaper prices.
5. Small-scale farmers get to learn about new practices.

FUTURE PROJECTIONS AND WORRIES

The climate projections for India can now be generated by using the report that was developed by Cline (2007). Table 5.9 highlights the temperature changes by region and rainfall changes by season for the period 2020–50.

We have considered the base period as 2002/03 as our period of study is 1950–2003. We have considered the situation where there is a 1.5°C rise along with 5 per cent increase in precipitation. The aggregate changes are incorporated in Table 5.9.

Table 5.9 clearly tells us that there is a tendency for the overall temperature to increase. We find that in the months of July–September, the overall temperatures have a tendency to decline. This can be attributed to the rainfall in those months which would bring down the temperature. Also, in all the four regions, the temperatures decline from January–March to April–June only to fall steeply in July–September. After the monsoons, the temperature again picks up in the months of October–December. But this rise is not so prominent due to the winter season. Also, in the months of October–December, the temperature rise of south east and south west is less prominent than that of north east and north west. This is because the south east and south west have the influence of the seas.

The projected figures on precipitation clearly tell us that all the four regions would experience an increase. However, during the season of January–March, the regions of north east and south east would experience a negative precipitation. These projections have been carried out by considering the latitude–longitude ranges of the regions with those of the states.

Table 5.9 Future impacts of temperature and precipitation

	Precipitation (%)				Temperature rise (°C)			
	NE	NW	SE	SW	NE	NW	SE	SW
Jun–Mar	−7.3	5.4	−14.3	15.3	2.93	2.10	2.14	2.77
Apr–Jun	12.8	5.2	23.9	19.3	2.76	2.04	1.67	2.58
Jul–Sep	16.3	21.3	7.2	4.2	1.77	1.88	1.44	1.43
Oct–Dec	5.5	32.3	1.2	4.1	2.67	2.03	1.89	1.78

Notes: NE = north east; NW = north west; SE = south east; SW = south west.

CONCLUSION

There has been concern regarding agricultural growth in India since the mid-1990s (McKinsey and Lonergan 1998). The growth rate of agriculture during 1997–98 to 2004–05 was only around 1.6 per cent per annum. Yield growth also declined significantly in the 1990s. It is important to realize the full potential of agriculture not only for higher overall growth but also to reduce poverty in India. Many policy issues are being discussed in this regard.

The analysis in this chapter clearly tells us that:

1. In a dynamic scenario, the impacts of climatic changes have an upward trend.
2. These changes clearly speak in favour of climatic sensitivity of Indian agriculture, especially with respect to the basic food items of rice and wheat.
3. Spatial autocorrelation becomes an important test of analysis and more so for Indian agriculture.
4. If the autocorrelation effects are introduced in the model, the intensity of the climatic sensitivity on Indian agriculture is much lower.

It has been observed that autocorrelation effects are present due to the continuous flow of information from the big farmers to the smaller ones. The better and the more numerous the information channels, the more prominent and visible would be such effects. However, policymakers have to answer the question of adaptation – how and when to adapt the appropriate policy.

A fresh look at the priorities of the Indian agricultural research system is necessary in light of the emerging prospects. It is known that India spends only 0.5 per cent of GDP on agricultural research as compared to more than 1 per cent in other developing countries. There is considerable potential for raising the effectiveness of these outlays by reordering the priorities in agricultural research and extension (Parikh and Gokarn 1992) without neglecting energy issues (Reddy and Parikh 1997). The returns to investment on research and extension will be much higher than those to other investments.

Thus public investment in agriculture, particularly in irrigation and research and extension, is needed for higher growth. There seems to be some trade-off between input subsidies and public investment in agriculture. The problem of mounting subsidies and their effect in terms of crowding out public agricultural investment has been highlighted in the Tenth Plan document. Input subsidies (on power, fertilizers, and irrigation) have been rising while public investment has been declining. Therefore reduction in input subsidies is important for raising public investment. Investment in rural infrastructure is more important for agricultural growth (Subramanian 2006) and rural development than trade liberalization per se. There has to be greater focus on dry land marginal areas for higher returns.

The speed of adjustment might be used to refer to the rate at which adaptation procedures are selected and implemented. Easy procedures take a relatively lesser period of time compared to the difficult ones which take longer. These adjustments should be implemented well before time. Moreover, the cost benefit analysis of each adjustment must be recorded before the adjustments are implemented.

Adaptation or adjustment policies cannot be devoid of the development plans. In other words, the policymakers must frame the programmes in such a way as to ensure that they match with the development ideas and desires of that region. Thus, these adaptation programmes will not be successful if the common man is not made a part of them. Rather, these programmes will be successful if the local people are made a part of them so that they truly become 'of the people, by the people, for the people'.

The reforms in the industrial sector initiated in 1991 benefited agriculture by shifting the terms of trade in favour of agriculture. However, the full benefits for agriculture can be expected to materialize when reforms directly affecting agriculture, such as removal of restrictions on domestic trade and processing, stepping up public investment in agriculture, and reforms in the management of infrastructure, are put in place.

The shift in consumption patterns also indicates that there is a need for crop diversification and improvements in allied activities like livestock and fisheries. This will also improve employment opportunities. Similarly, reforms should be undertaken for the growth of the rural non-farm sector. The rigidities in rural non-farm employment have to be removed in order to bring about rural transformation. The very low educational levels in rural areas limit access to better-paid employment, leaving rural workers with low skills, low productivity, and, therefore, probably low wage jobs. Education, infrastructure, credit, etc. have to be improved in order to raise the productivity of rural non-farm employment. The policy changes in the food and agricultural sector and in the rural non-farm sector are expected to reduce rural poverty through rural transformation.

NOTE

1. See Mora (2012).

BIBLIOGRAPHY

Arrhenius, S. (1896), 'Effect of carbonic acid in the air and its effect on ground temperature', *Philosophical Magazine and Journal of Science*, Series 5, Volume 41, April, pp. 237–76, reproduced at http://www. globalwarmingart.com.

Asthana, V. (1995), 'Climate change in Asia', Asian Development Bank.

Atul, K. Jain and Rama Nemani (2011), 'Land cover and land use changes and its effects on carbon and nitrogen dynamics in South and South East Asian Region', Department of Atmospheric Sciences, University of Illinois, Urbana, IL and NASA Ames Research Center, Moffett Field, CA, ISPRS Archives XXXVIII-8/W3, Workshop Proceedings.

Ausubel, Jesse H. (1983), Annexe, *Report of Carbon Dioxide Assessment Committee*, National Research Council, Washington, DC: National Academy Press

Chamberlin, Thomas C. (1897), 'The influence of great epochs of limestone formation upon the constitution of the atmosphere', *Journal of Geology*, **6**, 609–21.

Chandna, P., J.K. Ladha, M.S. Gill and Raj Gupta (2010), 'Impact of climate change on productivity of rice–wheat cropping system in the Indo-Gangetic Plains', Centre for the Study of Regional Development, Jawaharlal Nehru University, New Delhi, ISPRS Archives XXXVIII-8/W3, Workshop Proceedings.

Cline, W. (2007), *Global Warming and Agriculture: Impact Estimates by Country*, Washington, DC: Peterson Institute.

Coope, G.R. (1971), 'Fossil coleoptera as indicators of climatic fluctuations during the last glaciation in Britain', *Palaeogeography, Palaeoclimatology, Palaeoecology*, **10**, 87–101.

Dev, S. Mahendra (2007), 'Agricultural growth', in Kaushik Basu (ed.), *The Oxford Companion to Economics in India*, Delhi and Oxford: Oxford University Press, pp. 11–15.

Devi, Indira and G.S.L.H.V. Prasada Rao (2009), 'Economic impacts of climate variability and resource management options in crop production', Centre for Climate Change, Kerala Agricultural University, Thrissur, India, ISPRS Archives XXXVIII-8/W3, Workshop Proceedings.

Folland, C.K., D.E. Parker and F.E. Kates (1984), 'Worldwide marine temperature fluctuations 1856–1981', *Nature*, **310** (5979), 670–73.

Fourier, Joseph (1827), 'Mémoire sur les températures du globe terrestre et des espaces planétaires', *Mémoires de l'Académie Royale des Sciences*, **7**, 569–604.

Friss, E. and K. Lassen (1991), 'Length of the solar cycle – an indicator of solar activity closely associated with climate', *Science*, New Series, **254** (5032), 698–700.

IPCC (Intergovernmental Panel for Climate Change) (1992), *The First Assessment Report*, Cambridge: Cambridge University Press.

IPCC (1996), *Climate Change 1995, Vol 1 of The Science of Climate Change, IPCC Second Assessment*, Cambridge: Cambridge University Press.

IPCC (2001), *The Third Assessment Report*, Cambridge: Cambridge University Press.

IPCC (2007), *The Fourth Assessment Report*, Cambridge: Cambridge University Press.

James, I.N. (1994), *Introduction to Circulating Atmospheres*, Cambridge: Cambridge University Press.

Kiran, G. Sandhya and Shah Kinnary (2009), 'Remote sensing in carbon dynamics of cropland', Eco-Physiology RS-GIS lab, Department of Botany, M.S. University of Baroda, Gujarat, India, ISPRS Archives XXXVIII-8/W3, Workshop Proceedings.

Kumar, T.V. Lakshmi (2009), 'Climate change impact analysis on crop growing periods and rice yields over India', Department of Physics, Faculty of Engineering, SRM University, Chennai, India, ISPRS Archives XXXVIII-8/W3, Workshop Proceedings.

Kumar, K.S. and Jyoti Parikh (1998), 'Climate change impacts on Indian agriculture: the Ricardian approach', World Bank Technical Paper No. 402.

Kumar, K.S. and J. Parikh (2001), 'Indian agriculture and climate sensitivity', *Global Environmental Change*, **11** (2), 147–54.

Lal, R. et al. (1998), *The Potential of U.S. Cropland to Sequester Carbon and Mitigate the Greenhouse Effect*, Chelsea, MI: Ann Arbor Press.

Laut, P. and J. Gundermann (1998), 'Solar cycle length hypothesis', *Journal of Atmospheric and Solar-Terrestrial Physics*, **60**, 1719–28.

Lockwood, M. et al. (1999), 'A doubling of the sun's coronal magnetic field during the last 100 years', *Nature*, **399**, 3 June, 437–9.

Mall, R.K. et al. (2006) 'Water resources and climate change: an Indian perspective', *Current Science*, **90** (12), 25 June, 1610–26.

Manne, A.S. R.G. and R.G. Richels (1993), *Buying Greenhouse Insurance*, Cambridge, MA: MIT Press.

McKinsey, J. and S. Lonergan (1998), 'Measuring the impact of climate change on Indian agriculture', Technical Paper 402, World Bank, Washington, DC.

Mendelsohn, R., W. Nordhaus and D. Shaw (1994), 'The impact of global warming on agriculture: a Ricardian analysis', *American Economic Review*, **84** (4), 753–71.

Milesi, Cristina and Arindam Samanta (2010), 'Deteriorating food security in the semi-arid tropics', Division of Science and Environmental Policy, California State University Monterey Bay, NASA Ames Research Center, Moffett Field, CA and USA Department of Geography and Environment, Boston University, MA, ISPRS Archives XXXVIII-8/W3, Workshop Proceedings.

Misra, S.K. and V.K. Puri (2010), *Indian Economy*, 28th edition, Mumbai: Himalaya Publishing House.

Mitra, A.P. (ed.) (1992), 'Global change: greenhouse gas emissions in India – 1991', Methane Campaign, Scientific Report No. 2, NPL, Publication and Information Directorate, CSIR, New Delhi, India, June.

Mongia, Puran (2007), 'Global warming', in Kaushik Basu (ed.), *The Oxford Companion to Economics in India*, Oxford University Press.

Mora, R. (2012), 'The random effects estimator – quantitative microeconomics', Department of Economics, Universidad Carlos III de Madrid.

Murthy, N.S., Manoj Panda and Kirit Parikh (2000), 'CO2 emissions reduction strategies and economic development of India', IGIDR Discussion Paper.

National Accounts Statistics (various years), Central Statistical Organization, Government of India.

Oak Ridge National Laboratory (1993), *Trends '91 – A Compendium of Data on Global Changes*, Carbon Dioxide Information Analysis Center, Oak Ridge, TN: Oak Ridge National Laboratory.

Parikh, Jyoti and S. Gokarn (1993), 'Climate change and India's energy policy options: new perspectives on sectoral CO2 emissions and incremental costs', *Global Environmental Change*, September.

Parikh, Jyoti, and Kirit Parikh (1998), 'Free ride through delay: risk and accountability for climate change', *Environment and Development Economics*, **3** (3), pp. 347–409.

Pathak, H. (2009), 'Mitigation and adaptation to climate change in rice-wheat system with resource conserving technologies', Workshop Proceedings, Division of Environmental Sciences, Indian Agricultural Research Institute, Pusa, New Delhi, India, ISPRS Archives XXXVIII-8/W3

President's Scientific Advisory Committee (1965), 'Restoring the quality of our environment', Report of the Environmental Pollution Panel, available at http://dge.stanford.edu/labs/caldeiralab/Caldeira%20downloads/PSAC,%201965,%20Restoring%20the%20Quality%20of%20Our%20Environment.pdf.

Reddy, B.S. and Jyoti K. Parikh (1997), 'Economic and environmental impacts of demand side management programmes', *Energy Policy*, **25** (3), 349–56.

Reilly, J. (1996), 'Agriculture in a changing climate – impacts and adaptation', in IPCC (ed.), *Climate Change 1995*, Cambridge: Cambridge University Press.

Reilly, J.M. and D. Schimmelpfennig (1999): 'Agricultural impact assessment, vulnerability, and the scope for adaptation', *Climatic Change*, **43**, 745–88.

Revelle, Roger and Hans E. Suess (1957), 'Carbon dioxide exchange between atmosphere and ocean and the question of an increase of atmospheric CO_2 during the past decades', *Tellus*, **9**, 18–27.

Rowlands, M.J. (1992), 'Climate change, sea-level rise and the archaeological record', *Australian Archaeology*, **34**, 29–33.

Singh, U.K. and P.S. Salvekar (2009), 'Sea surface temperature over tropical Indian Ocean and south west monsoon performance', Indian Institute of Technology, Kanpur and Indian Institute of Tropical Meteorology, Pune, India, ISPRS Archives XXXVIII- 8/W3, Workshop Proceedings.

Stern, N. (2006), *The Economics of Climate Change*, The Stern Review, Cambridge: Cambridge University Press.

Subramanian, Arvind (2006), 'Growth experience', in Kaushik Basu, *The Oxford Companion to Economics in India*, Oxford University Press, pp. 231–6.

Sundaram, K. (2001), 'Employment and poverty in 1990s', *Economic and Political Weekly*, 11 August, 3039–49.

Sundararaman, N. (1997), *Reference Manual and Workbook of the IPCC 1996, Revised Guidelines for National Greenhouse Gas Inventories*, Geneva: IPCC.

Tyndall, John (1861), 'On the absorption and radiation of heat by gases and vapours', *Philosophical Magazine*, Series 4, **22**, 169–94, 273–85.

United Nations Environment Programme (2002), *Global Environment Outlook 3*, London and Sterling, VA: Earthscan.

University of Reading (2000), 'ECIFM: climate change and agriculture', http://www.ecifm.rdg.ac.uk/climate_change.htm.

Viskanta, R. et al. (1976), 'Modelling of the effects of pollutants and dispersion in urban atmospheres', Meteorology and Assessment Division Environmental Sciences, Research Laboratory, Research Triangle Park, North Carolina, February.

Wigley, T., R. Richels and J. Edmonds (1996), 'Economic and environmental choices in the stabilization of atmospheric CO_2 concentrations', *Nature*, **379**, 2402–33.

6. China's changing public health paradox and the new generation of health NGOs

Lai-Ha Chan and Ronald C. Keith

Non-government organizations (NGOs) that become involved in dissent in China run the serious risk of political suppression; however, many NGOs, as new 'social organizations' (*shehui tuanti*) have, in cooperating with the government, contributed to the successful achievement of practical policy goals within China's nascent civil society. Some have even been able to pursue 'comprehensive community intervention' that 'cuts across political, social and administrative boundaries' to further developing human rights.[1] This analysis specifically considers the role of China's health NGOs that have become active since the beginning of this century.

The new generation of health NGOs have provided additional resources and expertise to compensate for the inequalities in health service provision that have accompanied accelerated economic reform through the transition to the 'socialist market'. In terms of healthcare modernization China's market dynamic has spurred technological progress but the retreat of the state from public healthcare has become politically unacceptable especially in light of marked unevenness in the access to healthcare and new policy focus on social development that militates against GDPism as the exclusive focus on economic growth.

Chinese public healthcare policy has been and still is paradoxical in its strategy and outcomes. Under Mao's revolutionary leadership there was a strong emphasis on equal access to services and preventive medicine. In the transition to the socialist market, beginning in 1978 and especially since 1992, policy has focused on a market-oriented healthcare system, with the state playing a diminishing role in financing growing medical costs. The cost effectiveness of prevention was ignored while medicine and doctors focused on treating the problems of those who could afford healthcare.

In his day Mao Zedong bluntly focused everyone's attention on an all-important question: Who does the healthcare system serve? In Mao's time, the Chinese Communist Party (CCP)'s 'democratic centralism' and the mass line were the basis of organizational theory and practice. This was particularly manifested in health policy as well as the successful people's war that educated the masses in good sanitary practices and the quick identification of contagious diseases.

Mao challenged the exclusive focus on highly trained medical expertise in the cities and he was frustrated over the lack of resources available to China's vast rural population. Mao reprimanded the Ministry of Public Health in 1965:

> Tell the Ministry of Public Health that it only works for fifteen per cent of the total population of the country and that this fifteen per cent is mainly composed of gentlemen, while the broad masses of the peasants do not get any medical treatment. First they don't have any doctors; second they don't have any medicine. The Ministry of Public Health is not a Ministry of Public Health for the people, so why not change its name to the Ministry of Urban Health, the Ministry of Gentlemen's Health, or even the Ministry of Urban Gentlemen's Health. (cited in Schram 1974, p. 232)

If Mao forced 'redness' together with 'expertise', his policy did focus on universal access to healthcare. His thinking reflected the scarcity of central funding and the low standard of China's modern economic development. The rise of the 'barefoot doctor'[2] symbolized the importance of primary healthcare for the masses. In 1974, there were roughly one million 'barefoot doctors' in China. With the provision of basic medical care for most people in the urban and rural areas, China's healthcare system in Mao's China was comparatively more equitable and preventive than it is at present. Built on a network of 'barefoot doctors' and healthcare workers with rudimentary training acquired at the commune level, China's public health system provided more than 90 per cent of China's population with modest, but life-long government-subsidized healthcare. Various public projects and educational campaigns were carried out, aimed at improving hygiene and preventing disease outbreaks (Chan 2006).

Mass mobilization and related health education had proven to be a remarkably low cost solution to spreading diseases such as schistosomiasis. Mao had joyously penned a poem in 1958, 'Farewell, God of Plagues' to eulogize the success of controlling the disease (*The Economist* 2009). Not only did the system largely reduce the incidence of communicable diseases, it also significantly raised life expectancy from 35 years in 1952 to 68 years in 1985. Infant mortality dropped from about 250 per 1000 to 34 per 1000 live births during the same period (Liu 2004). Mao's public health strategy was praised as a model for the Third World by the World Health Organization (World Health Organization 2000, p. 14).

However, under Deng Xiaoping's economic reforms beginning in the late 1970s, which disbanded both the communes and the associated cooperative medical scheme, China's public health system switched to a user-pay, market-oriented system. The state cut back healthcare funding in line with the strategy suggested, for example, in the original Hainan economic model calling for 'small government, big society' (*xiao zhengfu, da shehui*). Individuals, popular organizations and state enterprise were invited to fill the gap left by a deliberately receding central government (Keith and Lin 2001, pp. 239–40).

While there has been spectacular progress in Chinese medical technology and medical skill as a result of reforms, there was a political failure to sustain the earlier focus on universality and prevention. Under the user-pay health system, people's basic health rights are no longer guaranteed. In many rural areas, public health services almost collapsed. The emphasis was no longer on taking healthcare to the masses in the countryside. Hospitals took on a much greater role in providing healthcare as primary healthcare declined.[3]

One of the biggest problems facing China's healthcare system today is the inequality in health outcomes. Private health (out-of-pocket) payments increased from 20 per cent in 1978 to nearly 60 per cent in 2000,[4] making health services increasingly unaffordable for most citizens. Furthermore, the bulk of medical resources is spent in urban areas, particularly in the coastal regions of Beijing, Shanghai, Zhejiang and Jiangsu (Kaufman 2010, p. 291).

According to the 2006 National Health Survey, conducted by the Ministry of Health, 48.9 per cent of those who needed medical care did not seek care from a doctor because of the cost involved, and 29.6 per cent of those who needed to be hospitalized could not afford it (Liu 2008). Overall, more than 90 per cent of the population were not satisfied with the health system.[5] These inequalities in health services and its poor public health system capacity have laid the ground for NGO participation in China's healthcare development. In addition, the present health system is by and large a system that solves the immediate medical problems of an elite base of paying customers. Preventative healthcare declined with the decline of primary healthcare. As a consequence, China's health system has struggled with (re-)emerging infectious diseases. Hepatitis, tuberculosis, human immunodeficiency virus infection/acquired immunodeficiency syndrome (HIV/AIDS), the severe acute respiratory syndrome (SARS) and avian flu are cases in point (Chan 2011a, pp. 260–66).

The international community has long expressed concern about China's health situation. At the end of the twentieth century, for example, there were many critical reports about China's 'AIDS villages' in Henan.

International NGOs have often claimed that China under-reports the seriousness of its HIV/AIDS problem (National Intelligence Council 2002). In a joint publication with China's Ministry of Health, the UN Theme Group on HIV/AIDS warned in its 1997 report, 'China responds to AIDS' that there was potential for a pandemic in China. There was concern that if the Chinese government did not respond effectively, the total number of HIV/AIDS cases in China could rise to ten million by 2010 (Chan 2011b, pp. 127–30).

To its credit, the Chinese government has learned from experience. The discovery of gaps in the local administration of health during the 2003 SARS crisis has reinforced new strategy and response to infectious disease.[6] The government has similarly realized that it needs the help of various sectors in order to combat its HIV/AIDS crisis. In June 2001 the Chinese government officially admitted the problem of the HIV/AIDS epidemic inside the country and called for a wide range of actors, including intergovernmental organizations (IGOs), NGOs, and government-spanning networks, as well as the private business sector, to work together to combat the disease inside China. This has directly fostered the development of health NGOs in China, particularly those focusing on the prevention and control of HIV/AIDS.

Even in the context of strict CCP leadership, NGOs may have a role to play, as suggested in the following conclusion in Feng-shi Wu's 2005 doctoral thesis: 'Evidence from China has shown that the politics generated by international non-state actors in a non-democratic country is complex, and despite strict macro-structures, transnational advocacy networks are emerging.'[7]

The context for NGO healthcare activity in China has, however, undergone important recent change as the result of a major reform started in 2009. During the Party leadership of Hu Jintao there has been what World Bank experts refer to as 'the re-engagement of the state' as China's leaders have recognized that 'growth is not enough'.[8]

Winnie Chi-man Yip and her colleagues' study of healthcare in China noted the return of once defunct communicable diseases along with unabated increases in non-communicable diseases. On the eve of the April 2009 reform '. . . the Chinese government was faced with widespread public discontent stemming from unaffordable access to healthcare, major financial risks associated with out-of-pocket medical expenses, and growing inequalities to access to healthcare and health status across regions and populations of different socio-economic status and between urban and rural areas'.[9]

Rather than reverting to Mao's emphasis on class struggle to promote social development Hu Jintao emphasized the traditional Chinese virtue of 'harmony'. Hu's 'harmony' has generally required more balance between economic growth and social development including the centrally funded provision of healthcare and educational opportunities. Mao had been long gone, when the 'God of Plagues' made an unexpected appearance in the midst of record economic growth and the question of 'urban gentlemen's health' remains pertinent. The frustrating gap in urban and rural access has been politically prioritized but it is still very much a problem.

In 2009, US$125 billion was assigned over a three-year period to bring about comprehensive universal healthcare. There have been impressive results in the expansion of healthcare coverage from 29.7 per cent in 2003 to 87.9 per cent in 2008 to 95.7 per cent in 2011, but there is considerable controversy over how meaningful this reform is. Despite new coverage many patients are still unable to meet the out-of-pocket costs of care.[10]

Especially pertinent to the present analysis is the 2009 reform emphasis on prevention, primary healthcare and the redistribution of financial and human resources to the poorer regions. The 2012 World Bank's report *China in 2030: Building a Modern Harmonious, and Creative High-Income Society* picked up on the new policy emphasis on harmony. The report, which did have considerable Chinese policy input, was possibly preaching to the converted when it called for a 'revitalised primary healthcare system' to play a key role in future healthcare coordination:

> Up to 85 percent of patients with non-communicable diseases need primary care because their conditions are best controlled with self-management. This care can – and should – be provided and managed by a strong primary health care system that is well coordinated with the broader health care system. Greater reliance on primary health care will mean less demand for hospital-based care, which has been a key factor behind the major cost escalation experienced in the Chinese health system in recent years.[11]

The Hu Jintao health reform re-prioritized both primary healthcare and prevention. The new reform emphasis on primary healthcare not only addressed the costs of the inefficient hospital-centric model that had developed in the first decades of reform, but also represented learning from the SARS experience when there was a serious possibility of the disease spreading rapidly throughout the countryside, where primary healthcare and the preventative approach had declined.

This chapter proceeds in three sections. It will first provide background to the emergence of health NGOs in China since the SARS outbreak of 2002–03, focusing on why the Chinese government has allowed the growth of health NGOs inside the country. The second section analyses the political challenges facing the new generation of health NGOs. Four different case studies analyse the rhetoric and reality of Beijing's ambivalent commitment towards particular types of NGO activities. The final section draws a conclusion on NGOs within the changing paradoxical context of Chinese healthcare development.

SARS AND EVOLVING HEALTH NGOS IN CHINA

The leadership of Hu Jintao and Wen Jiabao assumed power in the midst of the SARS outbreak in the spring of 2003. At the same time as it drew on past mass-line experience to wage a people's war on SARS, it ultimately expanded its organizational strategy to take into account the potential contribution of health NGOs. This change in attitude fostered the growth of NGOs (Thompson and Lu 2006).

The Chinese government, particularly the Ministry of Health and Guangdong officials, initially downplayed the severity of the disease and suppressed the release of information about the outbreak, while the virus spread out from southern China to other provinces in China as well as to overseas countries. China's belated response was severely criticized by the international community.

In a Politburo meeting on 17 April 2003, Hu Jintao and Wen Jiabao announced heightened determination to fight the disease (Chiu and Galbraith 2004, p. xxii). China wanted to allay international concerns over the pandemic's adverse impact on the world by showing that it was becoming more transparent in disclosing information about the disease at home. Since then, the central government has displayed a stronger commitment to improving public health, allowing the promotion of health NGOs inside the country.

The Chinese government was fully aware that in order to respond to its health challenges promptly and to effectively prevent and control HIV/AIDS and other infectious diseases, it needed help and cooperation from a wider range of popular organizations within civil society. To that end, the government allowed the growth of health NGOs. For example, in March 2005, Wu Yi, then Vice Premier and Health Minister, urged private companies and NGOs to play a greater role in halting the spread of HIV/AIDS in China. She indicated that the government could not win the battle by fighting the war alone (Pottinger 2005; Kahn 2005).

In addition, domestically, China has experienced a quiet sexual revolution in the context of its market-oriented economic reform. An ethnographic study of HIV/AIDS in China has showed a triangular relationship among economic growth, sex workers and HIV/AIDS (Hyde 2007). On the one hand, there is a plethora of new television and radio programmes on dating and relationship problems. Some members of the younger generation are more open about sex. Other segments of the population cling to traditional culture that treats sex as a taboo subject not to be aired in public. The social stigmatization of HIV/AIDS patients still encourages people to deliberately hide their real situation.

It is, therefore, hard for the government to identify the HIV/AIDS demography in the country. In addition, the fruits of China's economic reforms are highly concentrated in coastal cities. Approximately 150 million migrant workers from rural areas in inland provinces reside and work in major coastal cities. A direct link between floating migration and the diffusion of sexually transmitted diseases (STDs), including HIV/AIDS, has been identified.

Floating migrants are more likely to engage in high-risk behaviours, such as unprotected sex with multiple partners, needle sharing and drug injection (Yang 2004). For example, among all the transmission modes of HIV/AIDS in China, sexual transmission and intravenous drug use (IDU) accounted for 90 per cent of HIV/AIDS cases at the end of 2007.[12] As China turned outwards to the world economy, 'big society' was supposed to compensate for the receding state, or 'small government'. The latter sought new partnerships in society and this included new coordination with NGO activity in the socially and politically sensitive areas of HIV/AIDS.

According to the Ministry of Civil Affairs' official figures, in 1988 there were only 4446 registered non-profit NGOs in the country. This number quickly grew to 200 000 by 2001. By the end of 2009, more than 431 000 NGOs had registered with the Chinese government. An overwhelming majority (99.5 per cent) of the NGOs are operational at the provincial or local levels (National Bureau of Statistics of China 2010, p. 905). According to the most recent figure (Cheng and Shan 2011) about 400 non-profit grassroots organizations were actively involved in HIV/AIDS programmes in the country at the end of 2009.

All NGOs in China are required to register with the MCA or the State Administration of Industry and Commerce (*China Development Brief* 2007).[13] Those that do not register with the government agencies are normally called 'grassroots organizations'. The Chinese government has always maintained influence of varying degrees on recognized NGOs inside the country. Some of China's NGOs have been labelled in the West as GONGOs (government-organized NGOs). Government officials or retired high-ranking government officials often hold leadership positions in the GONGOs. Sovereignty and the supreme rule of the CCP remain the primary considerations in the government's treatment of the non-profit sector. Thus the Chinese government has allowed NGOs to develop in China, but only in so far as they do not constitute a challenge to China's regime.

POLITICAL CHALLENGES FACING HEALTH NGOS IN CHINA

Although the deteriorating healthcare system since economic reform and the subsequent SARS outbreak fostered the development of NGOs, particularly health NGOs, inside China, the development of NGO activity has been restricted by various governmental management policies.

First of all, in order to prevent any NGOs from growing so powerful that they could potentially threaten the CCP's supreme authority and control over society, all NGOs are prohibited from establishing regional branches. Unlike the conventional mass associations such as the Women's Federation, their operations are confined to one administrative level, such as county-level, provincial-level or national-level in Beijing. For instance, a provincial-level grassroots NGO cannot set up any branch outside the province wherein it is registered. To extend its work into other locations, it must register as an independent NGO in the new location. Although NGOs with the same objectives and conducting the same activities can exist simultaneously at all administrative levels, they have to remain as separate organizations. In addition, a new grassroots NGO cannot register if there is already an NGO doing similar work in the same administrative area. This policy has forced a lot of potential and capable grassroots NGOs to exist illegally and has directly hindered their extension (Lu 2007, pp. 3–4).

Owing to the government's tight regulations and its attempt to restrict the size of grassroots NGOs, most of them lack the organizational capacity to implement large-scale or even medium-scale activities. This is a vicious circle in that it leads government officials to believe that grassroots NGOs are not able to handle any major projects, such as the Global Fund's programme in fighting HIV/AIDS (as discussed below). In addition, the 'illegal' NGOs that fall outside the jurisdiction of the Ministry of Civil Affairs generally have little human and financial resources and connections at their disposal to provide quality service and to advocate policy changes. These combine to result in under-performance of many NGOs in both service delivery and policy advocacy in China (Lu 2007, 2009).

On the other hand, NGO-government relations are described as having 'dependent autonomy'. They often enjoy a high degree of *de facto* autonomy from the state's supervision while they rely on government agencies for patronage and/or financial resources in order to survive in an often inhospitable political environment. Paradoxically, mastery of the political skills in handling their complex and asymmetrical relationship with the state is not always helpful for them to discharge their duties properly to their clients. While cooperating with local government agencies, NGOs usually try to avoid making any challenges to the prevailing government policies. As a result, their 'voice' and services are only limited to those that are allowed by government and share the same 'voice' with it. The NGOs or activists who fought for the cause of the HIV/AIDS victims in Henan often faced an uphill battle due to the alleged complicity of Li Changchun, a member of the Standing Committee of the Politburo, in the infamous blood-selling scandal.[14]

More seriously, a deep concern over national sovereignty has led the Party to further tighten its control over NGOs and health activists. After the Falun Gong incident of 1992[15] and the 'colour revolutions' in some of the post-communist countries in the early 2000s, the Chinese government became concerned about the threat NGOs might pose to its authority and rule. As a result, it has strengthened its control over them and constricted their autonomy inside the country. The following analysis of four cases of NGO formation and development reveals the multi-faceted, if not contradictory, government attitude towards health NGOs and activists.

The Global Fund to Fight AIDS, Tuberculosis and Malaria

Despite the fact that China is rapidly emerging as a global economic power, it has been a main beneficiary of the Global Fund, a public-private partnership devoted to amassing and distributing financial resources to prevent and treat the three major infectious diseases of HIV/AIDS, tuberculosis and malaria. China began to engage the Global Fund in 2001.[16] Although the first two applications were rejected, China was finally granted almost US$283 million in the third round in 2004. This funding was largely devoted to the China CARES programme – a community-based HIV treatment, care and prevention programme in central China.[17] Its applications to the Global Fund were accepted in the following years. China has been thus far granted $946 million in total from the Global Fund and is the fourth-largest recipient of the organization, after Ethiopia, India, and Tanzania. As of March 2012, the Global Fund had disbursed a total of $606 million to China.[18]

In order to qualify for the Global Fund's financial support, China must meet certain criteria, of which one is to cooperate with grassroots NGOs inside the country. For example, in 2006, China pledged to use 43 per cent of the US$29 million it asked for to work with NGOs to combat the AIDS epidemic nationwide. Official figures show that between 2002 and 2006 China granted around US$2.5 million to fund 231 NGO projects to curtail the AIDS epidemic, covering almost all of its 31 provinces, municipalities and autonomous regions (*Kaiser Daily HIV/AIDS Report* 2006). However, the government's selection of NGOs for funding has been controversial. AIDS activists, such as Wang Yanhai, have criticized the system for its lack of transparency, its inequity and profligate 'double standards' (Dickie 2006).

On 26 November 2010, the Global Fund informed the Chinese National Centre for Disease Control and Prevention (China CDC) that the organization would freeze its $283 million payments to the Chinese government for its funding project, China CARES programme, 'due to a failure to implement the project', and 'lack of details regarding capacity building activities for civil society organisations' (*China Global Fund Watch Initiative* 2011). In May 2011, the Global Fund further announced that it would suspend all of its China programmes over suspected misuse of the money and the government's reluctance to engage with grassroots organizations in fighting HIV/AIDS.

One of the funding conditions for the China CARES programme is that the Chinese government must allocate at least 35 per cent of the project activity funds to civil society organizations (20 per cent in the first year, with progressive increases thereafter). However, according to progress reports and funding applications that China CDC submitted to the Global Fund in August 2010, while 753 civil society organizations applied for 8442 projects, the government only allocated $3.2 million to support civil society organizations' activities in the prevention and control of HIV/AIDS, accounting for only 10.9 per cent of the entire project's budget.

In addition, according to Global Fund Watch, a non-government group, the documents that the China CDC submitted to the Global Fund lacked sufficient information about the selection process of civil society organizations and their nature. There were allegedly 'no definite mechanisms to guarantee transparency, openness, and efficiency' regarding capacity building activities, as well as hostility towards civil society organizations. Therefore, the Global Fund decided to freeze its payments and required the Chinese government to resubmit its management and evaluation plan (*China Global Fund Watch Initiative* 2011, pp. 2–3).

In response, Chinese officials countered that many civil society organizations in the country could not be trusted to use the Global Fund's money properly and therefore the government distributed the funding to GONGOs, rather than grassroots organizations. Hao Yang, deputy director of the Bureau of Disease Prevention and Control under the Ministry of Health bluntly said, 'Currently, we can't find any Chinese non-governmental organization working with HIV and AIDS intervention that is able to handle the job' (Cheng and Shan 2011). According to

Chinese regulations, in order to be suitably qualified to be sub-recipients of the funding, applicants (organizations) must have a minimum of five years' experience in handling HIV/AIDS. However, government sources reveal that the average lifespan of community-based organizations in the country is only one year (Cheng and Shan 2011).

AIDS activists disagreed with the official view. For example, Chang Kun, an AIDS activist in Xinjiang region, alleged that 'they [officials] view our campaigning as troublemaking . . . I have been campaigning for AIDS patients for seven years now, and I rarely see people getting any benefits from the Global Fund.' Those 'official NGOs' (or GONGOs) had routinely received more than half the grant's funding. The AIDS grassroots organization that he is heading in Xinjiang, was awarded merely $3000 in 2005 and even this modest funding had to be refunded to the government due to its apparently 'illegal' status (LaFraniere 2011).

After mutual negotiations between the Global Fund and the Chinese authorities, the Global Fund lifted the temporary funding suspension on disbursements in August 2011. The Global Fund claims that the issue has not yet been fully resolved and that it lifted 'the freeze on the ground to ensure that AIDS work in the country continues while the Fund works with government officials, United Nations agencies and private groups to address the Fund's concerns' (Global Fund Observer 2011). Shortly after the freeze, the Health Ministry agreed to set up an independent office to manage the funding allocated to civil society organizations. Its minister, Chen Zhu, attended a meeting with community-based AIDS working groups in late June 2011 in Kunming, Yunnan province (Global Fund Observer 2011). In August 2011, the China CDC also launched an appeal for applications from grassroots organizations for handling a large portion of the money from the Global Fund (Cheng and Shan 2011).

Beijing Aizhixing Institute of Health Education

Another example of the scope and quality of the state-NGO dynamic is the Beijing Aizhixing Institute of Health Education (or Love Knowledge Action), which aims to increase China's awareness about HIV/AIDS and to advocate the rights and interests of those infected with the disease in China. This NGO was established in 1994 but it was not until 2002 that it was officially registered as the Beijing Aizhixing Institute of Health Education. The organization's website is highly informative about the disease of HIV/AIDS in China.[19] It engages in AIDS education, research and mobilizing public attention to the problem of HIV/AIDS, particularly the impact on AIDS orphans and the gay community in various provinces in China. The organization also has prolific publications on the HIV/AIDS situation in China, ranging from analysing government policies to educating the public on the prevention and control of HIV/AIDS.

The Institute receives financial sponsorship from various institutions, including UNESCO, the UK Embassy in Beijing, the National Endowment for Democracy in the US, AIDS organizations in France, and Misereor in Germany. The organization also helped the government in promoting AIDS awareness. For example, the China–UK HIV/AIDS Prevention and Care Program sponsored the organization to print pamphlets for the public in Xinjiang. A representative of the Institute in Beijing indicated in 2007 that the organization's work is accepted and recognized by the government.[20] In commenting on China's AIDS policy, Wan Yanhai emphasized that while the government provided funding to build infrastructure and public education, the frontline work of educating high-risk groups about HIV/AIDS is left with the NGOs (Branigan 2008).

However, the Aizhixing Institute's relationship with the government is often shaky. Sometimes its activities are banned outright by the government, which is surprising given the significant assistance that this particular NGO has rendered to the central as well as local governments in combating HIV/AIDS. Members of the organization have often been harassed or even put in jail by the government. For example, the organization planned a six-day AIDS conference, which aimed to provide a platform for AIDS experts and NGOs to discuss AIDS and blood-transfusion issues, in Beijing in November 2006. Just two days before it was due to start, the head of the organization, Wan Yanhai (1963–),[21] arguably one of the most outspoken AIDS campaigners in China, was detained by Beijing security forces. The conference was eventually cancelled after organizers received a call from Wan (Fan 2006).

After that, the organization developed a more positive working relationship with the government. However, before the Olympic Games in Beijing in 2008, the organization's activities were again subjected to official scrutiny. In May 2007, its executive director, Hu Jia (1973–) and his wife, Zeng Jinyan (1983–), a prolific Internet blogger who was named by *Time* magazine as one of the 100 most influential people in the world, were barred from leaving the country for a trip to Europe. They were suspected of being a threat to state security. According to Hu, Chinese officials were worried that the couple would set off opposition to the Olympic Games in Beijing in 2008. Hu and Zeng were prepared to show their audience in Europe a 30-minute documentary about their house arrest between July 2006 and February 2007 (Hilsum 2007).[22] While calling on the government to improve the treatment for HIV/AIDS patients in China, Hu had also published articles that likened the CCP to the Mafia and that advocated greater autonomy for Tibet.

Hu was finally sent to jail on 3 April 2008 for three and a half years on charges of 'incitement to subversion of the state' (Branigan and Watts 2008; Cody 2008).[23] This verdict immediately triggered condemnation from the European Union and the United States. While the EU called on the Chinese government to release Hu and awarded him the European parliament's top human rights award, the Sakharov Prize for Freedom of Thought, the then US Secretary of State Condoleezza Rice described the verdict as 'deeply disturbing' (Ng 2008). Amnesty International also declared that Hu's case showed that the Chinese government had betrayed its commitments to improve its human rights before the 2008 Olympic Games (Branigan and Watts 2008).

In September 2008, the Beijing Municipal Taxation Bureau investigated the organization for unpaid tax. In May 2010, it was subject to taxation investigation again for unpaid business taxes since 2002, including the organization's overseas funding between 2002 and 2007 (Frontline Defenders 2010). Due to alleged increasing pressure from the Chinese government, Wan fled to the US in May 2010 (Wong 2010). According to Wan, 'taxation investigations on selected NGOs are now widely used as a strategy of harassment' (Asia Despatch 2011). Nevertheless, after Wan's departure for the US, the organization maintains a limited level of operations (Wan 2012). While one can presume that the government desperately needs the organization's assistance in managing its AIDS problem inside the country, the government has, nonetheless, set strict limits on its activities.

The Gandan Xiangzhao and the Beijing Yirenping Centre

The Internet forum known as the Gandan Xiangzhao,[24] which literally means 'liver and gall bladder taking care of each other', was established in 2001 to promote social networking for people living with hepatitis B. Hepatitis B can be seen as endemic in China, with an estimated 120 million people living with the disease in the country. According to the pre-2005 civil servant recruitment rules, people living with hepatitis B were barred from joining the civil service in China.

However, two high-profile cases in 2003 successfully pushed the government to amend their civil service recruitment rules. In April 2003, a university graduate, Zhou Yichao, having passed the required entrance examination, was rejected for a civil service position in Zhejiang province because he tested hepatitis B positive. Distraught over this rejection,

Zhou stabbed to death a local government official. His crime eventually attracted the death penalty.

In the same year, another university graduate, Zhang Xianzhu, encountered a similar situation. He was one of the top candidates in the entrance examination for a position in Wuhui City in Anhui province. He was also denied the job because of the disease. Unlike Zhou, Zhang posted his story at the Gandan Xiangzhao, which stirred up a lively debate among members of the Forum. With their support along with that of a law professor at Sichuan University, Zhang filed a case against the Wuhui City's Personnel Bureau in November 2003. It was deemed as 'the first [court] case of hepatitis B discrimination in China' (Xinhuanet 2003). Although the court rejected Zhang's constitutional argument of equal protection of rights, it supported Zhang on technical grounds (Fu 2010).

After the cases of Zhou and Zhang, the Gandan Xiangzhao initiated a campaign against hepatitis B discrimination. It collected 1611 signatures from its citizens and submitted an open letter, calling on the National People's Congress at the end of 2003 to abolish the rules that prevented hepatitis B carriers from joining the civil service (*Zhongan Wang* 2003). In response to mounting public pressure, the State Council eventually issued a new 'Common Standard of Medical Examination in Civil Servant Recruitment (Trial)', stipulating that hepatitis B carriers were no longer barred from joining the civil service in China from January 2005 (Fu 2010; *Zhongguo Fayuan Wang* 2006).

In 2006, not content with merely running an online forum, the organizers of the Gandan Xiangzhao established a separate entity, the Beijing Yirenping Centre,[25] to arrange activities for hepatitis B carriers, such as conferences, educational workshops, etc. Since its establishment, the organization has been actively promoting equality rights, and coordinating activities for hepatitis B virus (HBV) carriers. It shortly became one of the most influential civil society organizations in China in providing legal protection and promotion of rights to those infected with hepatitis B. The organization has a close relationship with the Aizhixing Institute (Fu 2010; Webster 2011, p. 689).

However, while the organization became quite influential, it too has faced an uphill battle during Beijing's periodic clampdowns on NGO activities. The Gandan Xiangzhao was ordered to shut down twice in 2007 and 2008, and the Yirenping was also raided in 2009 by the police which seized leaflets that advised patients of their legal rights. The police claimed that the Centre needed to apply for a publication permit in order to distribute such leaflets (Jacobs 2009; Schiller 2009).

Food and Health-related NGOs

Research on China's healthcare in the West has highlighted weaknesses in China's approach to non-communicable diseases and the risk factors associated with tobacco, food contamination, poor diet and counterfeit drugs.[26] Since 2007, for example, there has been a spate of food and health safety scandals involving Chinese goods, ranging from fake green peas, bleaching pistachios, melamine-tainted milk power, as well as excessive pesticide residues in vegetables, fruit, teas, domestic fungi, animal products and seafood. Food safety is closely related to people's lives and health. Since the dawn of this century, food and health-related NGOs have been emerging in China to raise consumer concerns about safety and to improve self-regulation in the food industry. A typical example is the creation in February 2002 of the Pesticide Eco-Alternatives Centre (PEAC) based in Kunming, Yunnan, southwest China. The PEAC is the sole Chinese NGO promoting reduced use of pesticide, the usage of alternative forms of pest control and sustainable development.[27] In 2010, the organization received the 'National Advanced Social Organization Award' from the Chinese Ministry of Civil Affairs, indicating that its work is accepted and recognized by the government.

The organization has also developed a strong collaborative network with international environmental health organizations, such as the Swedish Society for Nature Conservation, the Rockefeller Brothers Foundation, and the International POPs (Persistent Organic Pollutants) Elimination Network. It has produced many publications on environmental health education in China, in both Chinese and English, and has received regular funding support from the Rockefeller Brothers Foundation.[28] Perhaps because the organization mainly focuses on community training and capacity building of plant protection, it has been able to maintain a good working relationship with the government.

On the other hand, other grassroots NGOs or individual whistleblowers on food and health scandals are often given severe punishment by the Chinese authorities. An example is Tang Lin, a villager from Chongqing's Fengjie county. He launched an online chat group to advocate food safety after his one-year-old son died in August 2008 from milk powder contamination. He was sentenced in June 2010 to one year in a 're-education through labour' camp for 'posing a threat to public security by scare-mongering' (AsiaNews.it 2010).

Another victim, whose trial caught both domestic and international attention and commanded widespread sympathy, is Zhao Lianhai. When his son became ill from the tainted milk formula, Zhao Lianhai created a website to champion the parents' cause and helped them to pursue lawsuits against dairy companies. He was given a two-and-a-half-year sentence in November 2010 for 'inciting social disorder' (Jacobs 2010; Yu and Jiao 2010). In an apparent move to give in to both internal and external pressure, the Chinese government reportedly released him on medical parole at the end of December 2010 after he did not appeal against the verdict.[29]

Other food and health-related NGOs in China include Wang Hai's Consumer Protection Hotline, the China Consumers' Association, established in 1984, the China Condiment Industry Association, and the China National Food Industry Association. However, owing to various restrictions on the registration, operation and funding of genuine NGOs in China, they have played only a minimal role in safeguarding food safety. The China Consumers' Association receives and handles complaints from consumers about food products, but it is a government-run organization. Like their counterparts in the US, the Chinese trade associations can regulate their member enterprises to raise the quality standards of their products. Nevertheless, due to their small size, they serve more as a conduit of information and regulations between the government and businesses (Ellis and Turner 2008, p. 3).

To sum up, the four cases above suggest that while there is a new NGO engagement with China's embryonic civil society in various areas of health activity, there are unresolved issues of political interference and protest that challenge the development of health NGOs.

CONCLUSION: CHINA'S HEALTH PARADOX

The history of Chinese healthcare delivery is replete with contradiction and paradox. Even as Mao unleashed the destructive forces of the Cultural Revolution he extended access to primary healthcare through the barefoot doctor programme at health stations in the countryside.

Subsequent economic reform reversed Mao's policy emphasis on prevention and universality, creating yet another paradox. In the late 1980s and 1990s, the state retreated before the introduction of the market and the commodification of health services filliped the development of medical technology even while it undercut universality, and prevention. Central funding declined and healthcare was largely left to provincial governments that lacked dedicated budget and capacity to ensure universal access to services, and this decentralization of costs remains one of the most pressing problems of healthcare reform.

It was in such a conflicted context that the modern Chinese health NGO found a new niche for itself. The Party's own mass associations used to monopolize their respective spheres of interest articulation in society;

however, they are increasingly sharing their responsibilities with NGOs and GONGOs within a new division of labour orchestrated by the state. There is now a huge new structure of popular organization in China as the CCP attempts to extend its leadership over new interests in society and to extend its control over the open spaces in society that have grown with market formation.

The third Plenary Session of the 16th Central Committee in October 2003 approved the 'Decision of the CCPCC (Chinese Communist Party Central Committee) on Some Issues Concerning the Improvement of the Socialist Market Economy', in which the CCP introduced the so-called 'five balances'. 'Balancing economic and social development' is one of them (*China Daily* 2003). The Decision was critical. It clearly recognized that there was a growing problem in the provision of healthcare as a basic element of 'social development'. Accelerated transition to the market to support ever increasing growth had led to a reduction of government support for primary healthcare.

Unexpected costs of serious hospital care and treatment can wreck the generational savings of an entire family. Those who are not wealthy cannot even afford prescriptions, for which there is a growing black market. The Party readily understood this situation as a challenge to political stability and its own legitimacy. The SARS experience pointed out the need to re-connect with the population and to bring the 'state back in' in terms of increased state health funding and of re-acquiring the lessons of mass populism in supporting preventive health and community education against contagious diseases.

The World Bank's *China in 2030* recently endorsed Hu's 'harmony' but stressed that primary healthcare would not only have to be revitalized, but that it would have to coordinate more effectively with higher level health units providing more rational gate-keeping and referrals within the national health system. While favouring neo-liberal principles in scaling back state enterprise, the Report emphatically recommended 'an increasing role for general revenues in financing healthcare, especially for the poor'.[30]

NGOs have achieved a beach-head in China's healthcare. They have brought new resources and expertise into China but they are regarded as only one aspect of a wider programme of organization. Their activity is based on sufferance, and now they will have to cooperate with a 're-engaged state' that is ready and willing to halt the retreat to 'small government' and to re-invest general revenues in primary healthcare and prevention.

There is of course one final paradox that reflects the Chinese approach to health in the era of globalization. China continues to tread a fine line between the quest for national autonomy and the need to ask for international assistance and to participate in international organization. In order to reflect the interaction of a new range of actors in the era of globalization, particularly on health issues, China has utilized multilateralism instrumentally. It has had to relax some of its stringent control over society. Hence, the Chinese government openly admits the problems and its health challenges, such as HIV/AIDS, and allows the involvement of global civil society in the campaign to combat the disease. The AIDS experience suggests the increasing importance of foreign funding and advice, but this experience has underlined political issues of control that loop back to the nature of the Party-State and its continued exercise of democratic centralism.

China is often at pains to stress the state-led nature of new international cooperation programmes. At the same time three years of post-2009 reform strongly suggest the 're-engagement of the state' in Chinese public health. NGOs have been constantly reminded that their presence in China depends on good behaviour, and this is not likely to change. The Party leadership is still inclined to discipline prominent health activists and NGOs who encourage dissent and direct criticism of China in the outside world. While China has become more integrated with the world economy and international organization, China persists in adopting a state-led multilateral approach to all NGOs. New social organization has proliferated outside the formally exclusive monopolies of the Party's mass associations. Health NGOs have been allowed to come through China's 'open door' but their entry is conditional on respect for Chinese sovereignty and Party control over the agenda for health delivery.

NOTES

1. See Ronald C. Keith and Zhiqiu Lin (2003).
2. In order to increase the number of medical practitioners in rural areas, Mao proposed that 'production teams' in the villages could recommend some peasants' children to receive short-term medical training courses at medical schools at the provincial level. After the training, they were sent back to villages as medical practitioners to provide basic medical services. Peasants called them 'barefoot doctors'. See Lai-Ha Chan (2006).
3. In the *World Health Report 2000 – Health Systems: Improving Performance*, published by the World Health Organization in 2000, China's healthcare system was rated poorly. Although not long ago still an admired model for developing countries, it was ranked 188th on the dimension of fairness of financial contribution, placing China as one of the most unfair countries in the world. See World Health Organization (2000).
4. *Zhongguo tongji nianjian* (*China Statistical Yearbook*), 2001 in the chapter on culture, sports, and public health.
5. The survey was conducted by *China Youth Daily* (*Zhongguo qingnian bao*) in August 2005 with 733 people aged 30 or above in the country. See Tang (2005).
6. The view on the public management of health crises reflected the lessons learned during

SARS and was specifically reflected in the subsequent Law on Infectious Disease and in new special regulations on AIDS. See Ronald C. Keith and Zhiqiu Lin, (2007).

7. See Fengshi Wu (2005), Doctoral Dissertation, abstract.
8. See Gu Xin's scathing review of Jane Duckett, *The Chinese State's Retreat from Health: Policy and Politics of Retrenchment* (Abingdon, UK: Routledge, 2011), in *China Information*, **26** (1), 118. The imbalance between growth and equal opportunity is outlined in the Executive Summary of the World Bank's *China in 2030*, 2012. The Report on p. 91 of Part 1 reflects on the contrast between the government's role vis-à-vis business affairs and health: 'In contrast, in health, education, social security and other areas typically considered to require government involvement because of market failures, China's government had initially retreated. Traditional arrangements were abandoned and government spending lagged behind'.
9. See Yip et al. (2012).
10. Huang Yanzhong in his *Foreign Affairs* article 'Sick man of Asia: China's health crisis' criticizes emerging health inequities and the neglect of the poor and floating population, whereas Yip et al. (2012, pp. 833–42), express 'cautious optimism' about progress in contemporary health reform. The explanation for the differences in interpretation are explored in Chen and Xu (2012). On the second page of the Winnie Chi-man Yip's analysis it is conceded that primary healthcare has yet to play a gate-keeping role and 'No shift in the flow of patients from high-level institutions to primary healthcare facilities has been recorded'.
11. World Bank (2012), Part 1, 2012, p. 52.
12. *A Joint Assessment of HIV/AIDS Prevention, Treatment and Care in China (2007)* (Beijing: State Council AIDS Working Committee Office of PRC and UN Theme Group on AIDS in China, 1 December 2007), pp. 4–5.
13. Those which fail to meet the requirements of the Ministry of Civil Affairs register as 'business enterprises' with the State Administration of Industry and Commerce. See Lu (2009), p. 5.
14. Li Changchun worked in Henan as Vice Governor, Acting Governor, Governor, Deputy Party Secretary and Party Secretary between 1990 and 1998 when the blood-selling scandal occurred in the province.
15. On 25 April 1992, Falun Gong held a protest against the Chinese media criticism of the organization in Beijing. It shocked the Central government as it organized more than 10 000 of its followers to surround the headquarters of the state apparatus in central Beijing for 13 hours. This incident alerted the government that NGOs could challenge its authority and absolute control of the country. Since then, Falun Gong has been declared 'an evil cult' and the government has further strengthened all NGO registration and supervision. See Lu (2007), especially note 7.
16. It was speculated that an application to the Global Fund for a grant worth US$90 million prompted the Chinese government to revise its record of the population of HIV/AIDS sufferers in 2001. The first application was rejected in early 2002, due to its unreliable statistics and closed attitude towards its HIV/AIDS problem. The government submitted another application later that year. See Rosenthal (2002).
17. Further details of the programme can be seen at http://portfolio.theglobalfund.org/en/Grant/List/CHN, accessed 8 April 2012.
18. See The Global Fund website on grant portfolio, 'China: East Asia and Pacific', http://portfolio.theglobalfund.org/en/Country/Index/CHN, accessed 2 April 2012.
19. The organization's website is http://www.aizhi.co/ and its twitter page is http://twitter.com/#!/azxing, accessed 30 March 2012.
20. Personal interview with a representative of the Beijing Aizhixing Institute of Health Education in Beijing, China in January 2007.
21. Wan Yanhai is a former Health Ministry official and a leading AIDS activist in China.
22. One of the documentaries about their house arrest is available at http://rfaunplugged.wordpress.com/2008/04/03/china-hu-jia-sentenced-to-3-12-years/, accessed 9 April 2012.
23. Hu was released on 26 June 2011.

24. See its website at http://www.hbvhbv.com/, accessed 8 April 2012. As a phrase, *gandan xiangzhao* means showing great devotion to each other.
25. Yirenping stands for *gongyi* (public welfare), *ren'ai* (kindheartedness) and *pingdeng* (equality). See its website at http://www.yirenping.org/INdex.asp, accessed 8 April 2012).
26. See Chen (2012).
27. Its website is http://www.penchina.org/, accessed 7 April 2012.
28. For details of its funding support from Rockefeller, see http://www.rbf.org/grant/10957/pesticide-eco-alternatives-center-yunnan-thoughtful-action-2, accessed 7 April 2012.
29. The verdict even came under criticism from Hong Kong deputies to the National People's Congress and Hong Kong delegates to the Chinese People's Political Consultative Conference. 'China: milk scandal protester reported free', *New York Times* (2010); 'Melamine activist "thanks" government for support', *Global Times* (2010a); 'Official warns against interference', *Global Times* (2010b).
30. World Bank (2012), Part 1, p. 52.

BIBLIOGRAPHY

Asia Despatch (2011), 'Wan Yanhai's statement on Chinese government crackdown on NGOs', 19 January, http://www.asiadespatch.com/2011/01/wan-yanhais-statement-on-chinese-government-crackdown-on-ngos/, accessed 2 January 2012.

AsiaNews.it (2010), 'Father who lost child to melamine-tainted milk gets a year of hard labour', 7 July, http://www.asianews.it/news-en/Father-who-lost-child-to-melamine-tainted-milk-gets-a-year-of-hard-labour-18872.html, accessed 9 April 2012.

Branigan, Tania (2008), 'China: quiet sexual revolution forces Beijing to admit dangers of AIDS: meetings and more funds but patients seeking talks with Premier are detained', *Guardian*, 18 April.

Branigan, Tania and Jonathan Watts (2008), 'Beijing, human rights: Beijing court jails dissident on subversion charges', *Guardian*, 4 April.

Chan, Lai-Ha (2011a), 'Oscillating between Mao and Deng? The domestic – global nexus of China's public health reform', in Lai-Ha Chan, Gerald Chan and Fung Kwan (eds), *China at 60: Global – Local Interactions*, Singapore: World Scientific, pp. 255–81.

Chan, Lai-Ha (2011b), *China Engages Global Health Governance: Responsible Stakeholder or System-Transformer?*, New York: Palgrave Macmillan.

Chan, Lai-Ha (2006), 'The evolution of health governance in China: a case study of HIV/AIDS', *Griffith Asia Institute Regional Outlook*, vol. 8, Brisbane: Griffith University, http://www.griffith.edu.au/__data/assets/pdf_file/0019/18235/regional-outlook-volume-8.pdf.

Chen, Lincoln and Dong Xu (2012), 'Trends in China's reforms: the Rashomon effect', *Lancet*, **379**: 9818, 3–9 March, pp. 782–3.

Cheng, Yingqi and Shan Juan (2011), 'AIDS fund manager is hard to find', *China Daily*, 6 August.

China Daily (2003), 'Third Plenary Session of the 16th Central Committee Open', 11 October.

China Development Brief (2007), 'Editorial: "GONGOs" are here to stay, but need to reform and open up', 19 April, http://www.chinadevelopmentbrief.com/node/1071, accessed 2 January 2012.

China Global Fund Watch Initiative (2011), issue 14, March, p. 2, http://www.cgf-watch.org/files/pdf_e/China_Global_Fund_Watch_Newsletter__Issue_14_EN_Final.pdf, accessed 28 December 2011.

Chiu, William and Veronica Galbraith (2004), 'Calendar of events', in Christine Loh and Civic Exchange (eds), *At the Epicentre: Hong Kong and the SARS outbreak*, Hong Kong: Hong Kong University Press, pp. xv–xxvii.

Cody, Edward (2008), 'Chinese rights advocate gets 3.5 year prison term', *Washington Post*, 3 April.

Dickie, Mare (2006), 'China to receive AIDS grants despite dispute over supervisory board', *Financial Times*, 11 July.

Ellis, Linden J. and Jennifer L. Turner (2008), *Sowing the Seeds: Opportunities for US–China Cooperation on Food Safety*, Washington, DC: Woodrow Wilson international Center for Scholars.

Fan, Maureen (2006), 'AIDS activist is detained on eve of meeting in China', *Washington Post*, 26 November, p. A16.

Frontline Defenders (2010), 'China: ongoing harassment of Beijing Aizhixing Institute for Health Education', 27 May, http://www.frontlinedefenders.org/node/2544, accessed 2 January 2012.

Fu, Hualing (2010), 'From online mobilization to offline action: Yirenping and public interest litigation', http://ssrn.com/abstract=1730122, accessed 8 April 2012.

Global Fund Observer (2011), 'Global fund to resume disbursements for grants to China', issue 157, 2 September, http://www.aidspan.org/documents/gfo/GFO-Issue-157.htm, accessed 28 December 2011.

Global Times (2010a), 'Melamine activist "thanks" government for support', 29 December http://china.globaltimes.cn/society/2010-12/606541.html, accessed 9 April 2012.

Global Times (2010b), 'Official warns against interference', 30 December http://china.globaltimes.cn/society/2010-12/607081.html, accessed 9 April 2012.

Hilsum, Lindsey (2007), 'Those dangerous bloggers', *New Statesman*, 11 June, pp. 20–21.

Hyde, Sandra Teresa (2007), *Eating Spring Rice: The Cultural Politics of AIDS in Southwest China*, Berkeley, CA: University of California Press.

Jacobs, Andrew (2009), 'China to end required testing for hepatitis B', *New York Times*, 12 October.

Jacobs, Andrew (2010), 'China sentences activist in milk scandal to prison', *New York Times*, 10 November.

Kahn, Joseph (2005), 'World briefing Asia: China: business urged to help fight AIDS', *New York Times*, 19 March.

Kaiser Daily HIV/AIDS Report (2006), 'Chinese government to boost cooperation with support for NGOs fighting HIV/AIDS, Health Ministry official says', 5 October.

Kaufman, Joan (2010) 'Policy case study: public health', in William A. Joseph (ed.), *Politics in China: An Introduction*, New York: Oxford University Press.

Keith, Ronald and Lin Zhiqiu (2007), 'SARS in Chinese politics and law', *China Information*, November, 414–18.

Keith, Ronald C. and Lin Zhiqiu (2003), 'The making of a Chinese NGO: the research and intervention project on domestic violence', *Problems of Post Communism*, **50** (6), November–December, 45–8.

Keith, Ronald and Lin Zhiqiu (2001), *Law and Justice in China's New Marketplace*, New York: Palgrave.
LaFraniere, Sharon (2011), 'AIDS funds frozen for China in grant dispute', *New York Times*, 20 May.
Liu, Yating (2008), 'Kanbing nan, nan zai nali? Kanbing gui, gui zai hechu?' ('Getting health care is hard, why is it hard? Seeing a doctor is expensive, why is it expensive?'), *Zhongguo Shangjie* (*Business China*), (3), pp. 36–8.
Liu, Yuanli (2004), 'China's public healthcare system: facing the challenges', *Bulletin of the World Health Organization*, **82** (7), July, 532–8.
Lu, Yiyi (2007), 'NGOs in China: development dynamics and challenges', China Policy Institute Discussion Paper, No. 18, Nottingham: The University of Nottingham.
Lu, Yiyi (2009), *Non-Governmental Organisations in China*, Abingdon: Routledge.
National Bureau of Statistics of China (2010), *Zhongguo tongji nianjian 2010* (*China Statistics Yearbook 2010*), Beijing: China Statistics Press.
National Intelligence Council (2002), 'The next wave of HIV/AIDS: Nigeria, Ethiopia, Russia, India, China', (ICA 2002-04D, September), http://www.fas.org/irp/nic/hiv-aids.html, accessed 29 May 2012.
New York Times (2010), 'China: milk scandal protester reported free', 28 December, http://www.nytimes.com/2010/12/29/world/asia/web-briefs-Chinabrf.html, accessed 9 April 2012.
Ng, Tze-wei (2008), 'Activist denied final chance to appeal: lawyer Hu Jia "unavailable" on last day', *South China Morning Post*, 18 April.
Pottinger, Matt (2005), 'China urges businesses to stem AIDS', *Asian Wall Street Journal*, 21 March.
Rosenthal, Elisabeth (2002), 'China now set to make copies of AIDS drugs', *New York Times*, 7 September.
Schiller, Bill (2009), 'China tightens reins on dissent', *Toronto Star*, 8 August.
Schram, Stuart Reynolds (1974), *Mao Tse-Tung Unrehearsed: Talks and Letters, 1956–71*, Harmondsworth: Penguin.
Tang, Yonglin (2005), 'Yiliao gaige: jiucheng gongzhong bumanyi' ('Healthcare reform: 90% of the population were not satisfied'), Yi gai wang (Healthcare reform net), 23 August, http://yigai.org/NewsInfo.aspx?id=1, accessed 23 February 2011.
The Economist (2009), 'Hello again, God of Plague', 18 June.
Thompson, Drew and Xiaoqing Lu (2006), 'China's evolving civil society: from environment to health', *China Environment Series*, Woodrow Wilson International Centre for Scholars, No. 8, pp. 27–39.
Wan Yanhai (2012), 'The 18th anniversary of Beijing Aizhixing Institute', Voachineseblog, http://voachineseblog.com/wanyanhai/2012/03/20/18-years-of-beijing-aizhixing/, accessed 2 April 2012.
Webster, Timothy (2011), 'Ambivalence and activism: employment discrimination in China', *Vanderbilt Journal of Transnational Law*, **44**, 643–709.
Wong, Edward (2010) 'AIDS activist leaves China for US, citing pressure', *New York Times*, 10 May.
World Bank (2012), *Chin 2030: Building a Modern, Harmonious and Creative High-Income Society*, http://www.worldbank.org/contact/dam/Worldbank/document/china-2030-complete.pdf.
World Health Organization (2000), 'The World Health Report 2000 – health systems: improving performance, http://www.who.int/whr/2000/en/, accessed 9 April 2012.

Wu, Fengshi (2005), 'Double-mobilization: transnational advocacy networks for China's environment and public health', doctoral dissertation, University of Maryland, College Park, abstract.
Xinhuanet (2003), 'Quanguo meiti jujiao "yigan qishi"' ('All the media in the country focus on 'hepatitis B discrimination'), 15 November, http://big5.xinhuanet.com/gate/big5/news.xinhuanet.com/newscenter/2003-11/15/content_1179802.htm, accessed 8 July 2012.
Yang, Xiushi (2004), 'Temporary migration and the spread of STDs/HIV in China: is there a link?' *The International Migration Review*, **38** (1), 212–35.
Yip, Winnie Chi-man, Hsiao, William C. et al. (2012), 'Early appraisal of China's huge and complex health-care reforms', *Lancet*, **379**: 9818, 3–9 March, pp. 833–42.
Yu, Verna and Priscilla Jiao (2010), 'Outrage as tainted milk "hero" goes to jail', *South China Morning Post*, 11 November, through LexisNexis.
Zhongan Wang (2003), 'Yi 1.2 yi min yigan xiedaizhe minyi, 1611 gongmin tiqi songsu' ('1611 citizens initiate a litigation in the name of 120 million hepatitis B virus carriers'), 26 November, http://health.big5.anhuinews.com/system/2003/11/26/001198852.shtml, accessed 8 April 2012.
Zhongguo Fayuan Wang (2006), 'Zhang Xianzhu an zhongjie yigan qishi' ('The case of Zhang Xianzhu ended hepatitis B discrimination'), 3 November, http://rmfyb.chinacourt.org/public/detail.php?id=102369, accessed 8 April 2012.

7. Towards universal sanitation: the Cambodian case

Paul Howard

INTRODUCTION

The statistics related to inadequate sanitation globally are sobering. It is estimated that around 2.6 billion people on the planet do not have access to an improved sanitation facility (WSSCC 2012a). Of those 2.6 billion people, around 70 per cent live in rural areas and 1 billion still openly defecate (WSSCC 2012a). At the conclusion of the Water Week conference in Stockholm in August 2011, the Chair of UN Water stated United Nations (UN) Water's wish to set an explicit target of achieving 'universal provisioning of safe drinking water, adequate sanitation and modern energy services by the year 2030' (UNHABITAT 2011). More immediately though, one of the specific targets of the UN's Millennium Development Goals (MDGs) was to 'halve, by 2015, the proportion of people without sustainable access to safe drinking water and basic sanitation'. In regard to this target, a recent UN assessment of the progress in meeting the various MDGs determined that the first part of this two-part goal was on track to be exceeded with the percentage of people with access to clean drinking water well in excess of the target. Unfortunately though, the second part of the target, access to basic sanitation, was unlikely to be met by 2015.

In Cambodia, the picture is one that has been greatly defined by previous internal conflict, which has left the country economically disadvantaged with a relatively low standard of living and poor infrastructure. Inadequate sanitation facilities and practices may result in adverse impacts on human health, with the young particularly affected. Along with the significant community and individual health consequences of inadequate sanitation, there are also significant negative economic impacts. Whilst certain economic impacts may be more readily quantifiable, there are long term economic costs associated with poor sanitation access and practices. For example, lost productivity due to illness and students performing poorly due to health issues will have negative long term impacts on the economic development of the country.

Addressing Cambodia's low levels of sanitation access is essential for the country's economic and social development. Given the importance of this issue, there are two purposes to this chapter. The first is to provide a snapshot of the Cambodian situation in terms of access to sanitation in both quantitative and qualitative senses. It is hoped that the data and analysis presented in this chapter will help facilitate an improved understanding of Cambodia's sanitation performance. The percentage of Cambodians with access to an improved sanitation facility is extremely low, in both regional and global terms. Indeed, as is evident in data presented in this chapter, the percentage of Cambodians with access to improved sanitation is the lowest of any country in the Asian region. To put this in perspective, Cambodians are twice as likely to have a mobile phone as they are to have access to a toilet (De Laney 2011). Access to adequate sanitation facilities remains an enormous challenge, particularly outside of the major centres, and there exists a huge disparity between rural and urban levels of access to sanitation facilities. This rural and urban disparity in terms of access to adequate sanitation provides another degree of complexity to the issue.

The second purpose of this chapter is to highlight some of the actions being taken to improve access to sustainable sanitation, particularly within the local context, and to consider the integration between the various actors. I argue that it is vital that goals set by international organizations have reference to the realities of implementation of programmes at the local level. Within Cambodia, there is little doubt that many actors are making great strides within those local contexts. Many non-government organizations (NGOs) in Cambodia have been undertaking substantial fieldwork and implementing concrete plans within specific areas and villages to improve sanitation access. Whilst there are many positive examples of action by specific actors to address Cambodia's relatively poor level of sanitation coverage, it is not clear that integration of purpose and action exists among the actors concerned within the Cambodian context. Exactly how this integration will be achieved is not the immediate focus of this chapter. However, the achievement of universal access to sanitation for all Cambodians will require integration among and between the actors at all levels.

To achieve both of the above purposes, this chapter will be structured as follows. Firstly, the global and national context will be considered, specifically in terms of the UN MDGs and the Cambodian MDGs (CMDG), with the following section providing analysis of the relative performance of Cambodia with regard to access to sanitation in the broader Asian region. This includes some analysis of urban/rural disparities in a comparative regional context with the next section dealing with the urban/rural disparity within the Cambodian context. Having gained a clearer picture of the global, regional and national situation, the subsequent section provides examples of the implementation of initiatives at the local level.

THE GLOBAL AND THE NATIONAL

The Millennium Development Goals (MDGs), developed by the UN, are a set of eight major goals, with each goal having a number of defined targets. Among the targets attached to Goal 7 ('ensure environmental sustainability') is the water and sanitation specific target 7c, which is the stated aim to 'halve by 2015, the proportion of the population without sustainable access to safe drinking water and basic sanitation' (UN 2012). As of 2012, included in its assessment of progress toward this target the UN (2012) makes three main points with regard to the sanitation aspect:

1. 'With half the population of developing regions without sanitation, the 2015 target appears to be out of reach'
2. 'Disparities in urban and rural sanitation coverage remain daunting' and
3. 'Improvements in sanitation are bypassing the poor'

Unfortunately, in the Cambodian context, all three of these observations are applicable. Indeed, Cambodia has the lowest national percentage of people with access to improved sanitation in the region (UNESCAP 2011). With sanitation access as part of the broader MDGs, it was with the Cambodian national context in mind that the Cambodian Ministry of Planning (CMP 2010) developed the CMDG in 2003:

1. Eradicate extreme poverty and hunger
2. Achieve universal primary education
3. Promote gender equality and empower women
4. Reduce child mortality
5. Improve maternal health
6. Combat HIV/AIDS, malaria and other diseases
7. Ensure environmental sustainability
8. Develop a global partnership for development
9. De-mining, explosive remnants of war (ERW) and victim assistance

Of these CMDG, the most pertinent to this study is CMDG 7: 'Ensure Environmental Sustainability'. Whilst this CMDG is a broad, unquantified goal, it is disaggregated into five more specific measures, of which 7.1 is a specific sanitation measure (CMP 2010):

CMDG 7.1	Ratio of latrines to families
CMDG 7.2	Per cent of families using pipe or tube well water
CMDG 7.3	Per cent of families who filter or boil water
CMDG 7.4	Land conflicts per 1000 families
CMDG 7.5	Per cent of families using chemical fertilizer or pesticides

CMDG 4: 'Reduce Child Mortality' and CMDG 5: 'Improve Maternal Health' are also directly linked to the issue of sanitation, as poor sanitation affects health outcomes for entire communities.

In addition to the CMDGs, the Cambodian Government in conjunction with the UN also employ country specific measures for household poverty. Number 11 of the 13 indicators of 'Poverty' makes specific reference to the number of latrines per family (CMP 2010).

1. Household size (–)
2. Ratio of children 6–14 out of school (–)
3. Ratio of literate women 18–64 (+)
4. Ratio of men 18–64 to all (+)
5. Concrete house per family (+)
6. Thatch house per family (–)
7. Ratio of house with electricity (+)
8. TV per family (+)
9. Motorbike per family (+)
10. Bicycle per family (+)
11. No latrine per family (–)
12. Ratio of family use (–)
13. Ratio water in home less than 150m (+)

In a presentation, 'A toilet in every home: sanitation advocacy in Cambodia', the Deputy Prime Minister and Minister, Ministry of Rural Development, outlined the national situation for sanitation in rural Cambodia, with the explicit national goal that 'Every person in rural communities has sustained access to safe water supply and sanitation services and lives in a hygienic environment by 2025' (Cambodian Ministry of Rural Development 2008). To achieve this goal, the Minister said there would be incremental targets with the year 2015 (the target year for MDGs) seeing 30 per cent of rural people with access to improved/adequate sanitation. The UN definition of an 'improved sanitation facility' is 'one that hygienically separates human excreta from human contact' (WHO/UNICEF 2010).

There is great global variance between access to basic sanitation in rural and urban areas with open defecation (the most hazardous form

Table 7.1 The criteria of 'improved' versus 'unimproved' sanitation facilities as used by the Joint Monitoring Programmes by UNICEF and WHO

Improved sanitation options	Unimproved sanitation options
Flush or pour flush to sewerage	Any public or shared latrine
Flush or pour flush to septic tank or pit	Flush or pour flush to elsewhere
Pit latrine with slab	Open pit latrine without slab
Ventilated improved pit Latrine	Latrine overhanging water
	Other options

Source: WHO/UNICEF (2010).

of activity to human health outcomes) remaining pervasive, particularly in parts of sub-Saharan Africa and Asia. Inadequate and sub-standard sanitation facilities and practices may lead to dire health consequences for entire communities. In the International Year of Sanitation in 2008, the UN estimated that more than 5000 children under the age of five die every day as a result of diarrhoea (WSSCC 2012b), frequently the direct consequence of sub-standard or non-existent sanitation facilities. In Southeast Asia, there is still much room for further progress with only 79 per cent of urban dwellers and 59 per cent of rural dwellers, respectively, using an improved sanitation facility (UNESCAP 2011).

COMPARATIVE PERFORMANCE

Within the context of the broader Asian region, Cambodia's sanitation coverage is poor at both an overall national level and particularly in terms of access to sanitation in rural areas (Figure 7.1). For example, of the countries for which UNESCAP (United Nations Economic and Social Commission for Asia and the Pacific) data is available, Cambodia performs worst at the national level, with both India and Afghanistan, the next poorest performers, having higher rates of sanitation access at the aggregate national level (UNESCAP 2011). Whilst Cambodia does perform better than many other developing countries in the Asian region in terms of percentages of the urban population with access to improved sanitation, at the rural level, the picture is very different. Indeed, with only 18 per cent of rural people having access to an improved sanitation facility, Cambodia performs 3 percentage points worse than India and 12 percentage points worse than Afghanistan (UNESCAP 2011).

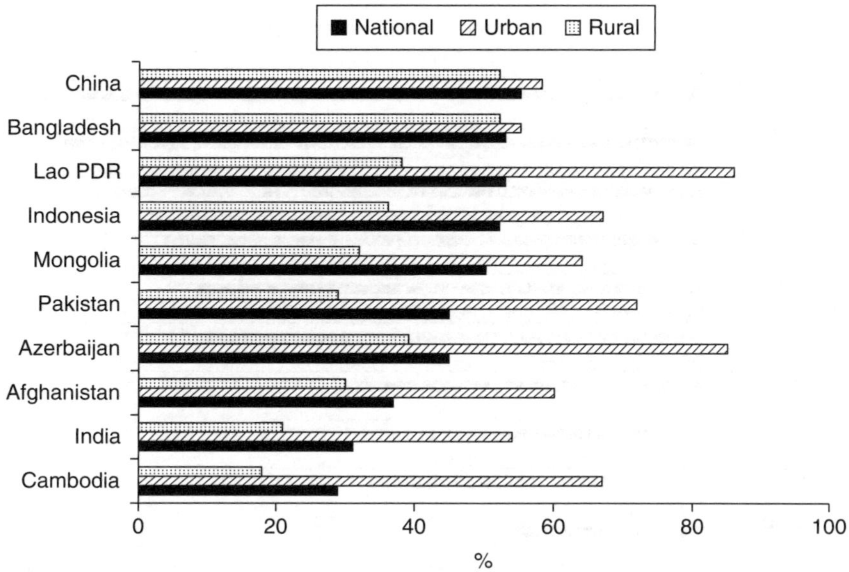

Source: UNESCAP (2011).

Figure 7.1 Access to improved sanitation for selected Asian countries with total national improved sanitation coverage of 55% or less

Cambodia's relative level of sanitation access is also poor in comparison to its Southeast Asian neighbours. On a national basis (urban and rural combined), Cambodia has just 29 per cent of the population with access to an improved sanitation facility, by far the lowest level in Southeast Asia (UNESCAP 2011). At the urban level, Cambodia, with 67 per cent of the population having access to improved sanitation, is equal to Indonesia but comes in below all other Southeast Asian states, such as Timor-Leste with 76 per cent coverage, the Philippines with 80 per cent coverage, Laos with 86 per cent coverage, Myanmar with 86 per cent coverage and Thailand with 96 per cent coverage (UNESCAP 2011).

Rural Southeast Asia

At the rural level, the data provides glaring evidence of Cambodia's relatively poor sanitation coverage. Only 18 per cent of rural Cambodians have access to an improved sanitation facility (UNESCAP 2011). The next lowest percentages of access to an improved sanitation facility in

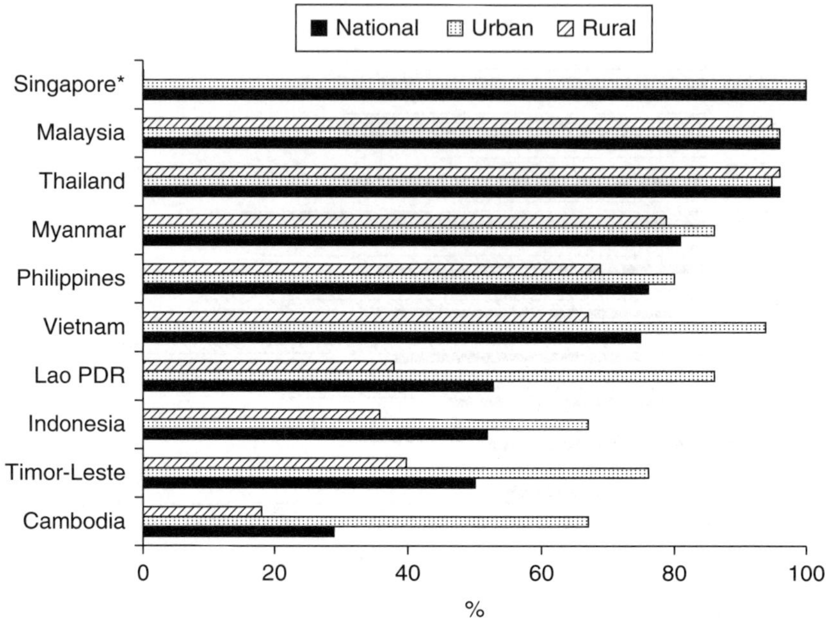

Note: * Data for Singapore only available as urban and national, that is, no rural component.

Source: UNESCAP (2011).

Figure 7.2 Access to improved sanitation in Southeast Asian countries (%)

rural areas in Southeast Asia are in Indonesia and Laos PDR where 36 per cent and 38 per cent respectively of the rural population have access to an improved sanitation facility (UNESCAP 2011), around double Cambodia's levels of rural sanitation access. Other South east Asian states perform far better at the rural level, with the next lowest percentages of access to improved sanitation facilities in rural areas being Vietnam which has 67 per cent rural sanitation access and the Philippines with 69 per cent of the rural population with access (UNESCAP 2011) (Figure 7.2).

NATIONAL URBAN/RURAL DISPARITY

At a national level, there is enormous disparity between urban and rural areas in terms of usage of sanitation facilities. For example, as of 2008,

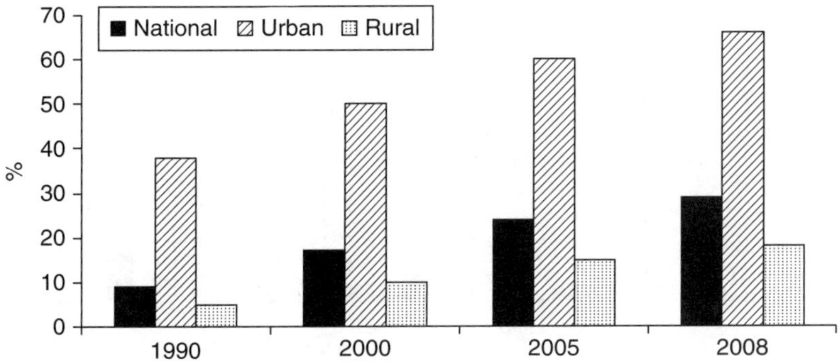

Source: UNESCAP (2011).

Figure 7.3 Access to improved sanitation in Cambodia, 1990–2008

67 per cent of the urban population used an 'improved' sanitation facility as opposed to only 18 per cent of the rural population (WHO/UNICEF 2010). The 18 per cent of the rural population with access, to an improved sanitation facility did increase from the 5 per cent of the population with access to an improved sanitation facility in 1990, a 260 per cent increase over the period. However, this compares to the urban area where 38 per cent of people had access to an improved sanitation facility in 1990 as compared to 67 per cent with access to an improved sanitation facility in 2008 (WHO/UNICEF 2010).

Therefore, whilst the rate of increase in percentage of people with access in rural areas was far higher, on the basis of the actual increase in the percentage of population with access, the urban area saw an increase of 29 per cent of the population with access, as opposed to an increase of 13 per cent of the rural population with access. In terms of the increase in the actual percentage of people with improved access, the rural areas have seen far less increase than the urban population (Figure 7.3).

The disparity between rural and urban populations is most starkly represented by the data pertaining to the practice of open defecation. For example, in rural areas, the number of people engaging in open defecation decreased from 89 per cent in 1990 to 75 per cent in 2008 (WHO/UNICEF 2010) (Figure 7.4). This compares to the urban areas where the percentage of people openly defecating fell from 48 per cent of the population in 1990 to 22 per cent of the population in 2008 (WHO/UNICEF 2010). (Figure 7.5). This means that rural dwellers are three times as likely to engage in open defecation as urban dwellers.

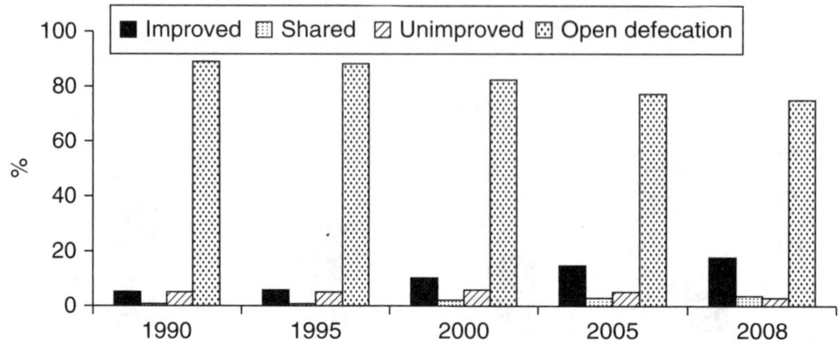

Source: WHO/UNICEF (2010).

Figure 7.4 Rural Cambodia sanitation facility use by type (%)

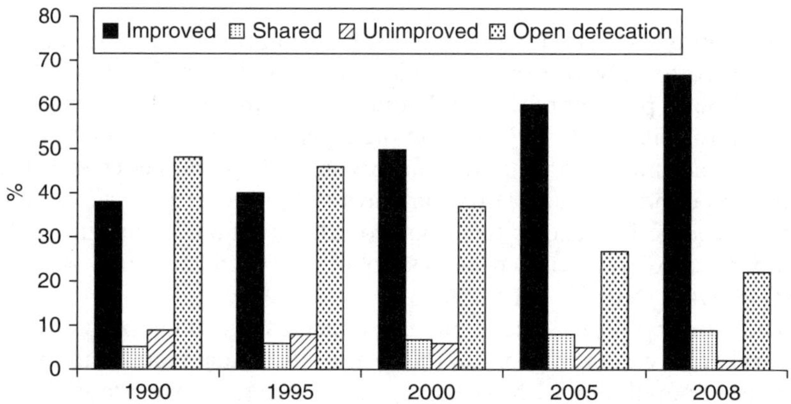

Source: WHO/UNICEF (2010).

Figure 7.5 Urban Cambodia sanitation facility use by type (%)

LOCAL CONTEXT

From the above discussion, it is clear that Cambodia's levels of sanitation access are poor on both a comparative regional basis and in terms of the national rural/urban disparity of sanitation access. To address these comparative disparities ultimately requires implementation within the local context. The need for integration between global, national and local level actions in addressing health issues is well documented. For example, in a study of the efficacy of organizational integration in the prevention of unsafe abortion, Zaidi et al. (2010) determined the need for comprehensive 'international collaboration' between and among actors at international, national and local level, describing it as a 'basic requirement for the success' of the initiative under study. The initiative involved a range of actors including international organizations such as the World Health Organization (WHO) and United Nations Population Fund (UNFPA) along with a range of NGOs operating within global, national and local contexts. With regard to the national context, one qualitative study (Ejaz et al. 2011) employed interviews to obtain a comprehensive picture of the partnership arrangements between government agencies, NGOs and donors in Pakistan. The authors point out the potential conflicts that exist at times in functional aspects of the NGO and public sector relationships, arguing that 'NGOs and public sector keep putting responsibilities over each other to do certain things and basically it results from a lack of clarity in the roles and responsibilities of each sector while within partnership and also when working separately' (Ejaz et al. 2011). Further to this, at the UN level, Ejaz et al. (2011) note that the achievement of the 2015 MDG targets related to 'quality and efficacy of medical care' will necessitate stronger 'collaboration' between government health programmes and relevant NGOs.[1]

In Cambodia, one interesting local level activity undertaken by the NGO WaterAid highlights the importance of grassroots activities and positive outcomes. 'Community Mapping' is utilized by WaterAid to engage entire local communities in understanding their respective local area sanitation infrastructure along with implementing the necessary actions to improve sanitation facilities (Wicken 2010). James Wicken of WaterAid (2010, p. 24) succinctly explains the process.

> One of the first steps in the process is a transect walk by a group from the community through the entire village to locate the areas used for open defecation and visit different types of latrines. Standing in the areas used for open defecation with a visitor to the community, inhaling the smells and asking questions about who defecates in the area is a very effective motivating tool. After this 'walk of disgust' the community makes a map on the ground showing households, resources (such as schools), open defecation areas and water points. The mapping gets all the community members involved in a practical and visual analysis of the community sanitation situation. The map is used to start a conversation about the flow of faeces from the open defecation areas to water sources and how this results in illness. Later in the process the map is transferred to paper by the community and used as a monitoring tool to check which households have constructed and use a toilet.

Obviously, implementing such programmes requires local knowledge of cultural context, without which misunderstandings or offense may be problematic. It is therefore a valuable example of the reality that project implementers and planners are faced with. Despite the seemingly straightforward approach, this form of mapping still involves considerable time and money. Wicken (2010, p. 24) cites the specific case of Malawi where it typically costs between $10 000 and $20 000 to map just one district, the cost dependent on variables such as the size of the district, the population density and the number of 'water points'. It takes around a month to collect the data and another two months for the required analysis and the provision of feedback to the respective district officials and interested parties (Wicken 2010, p. 25).

This is one example of the Community-Led Total Sanitation (CLTS) approach, pioneered in 1999 by independent Bangladesh development consultant Karmal Kar.[2] Synne Movik and Lyla Mehta (2010) of the STEPS Centre, Institute of Development Studies, University of Sussex, have asserted that the CLTS approach is the type of 'radical' approach that is needed in trying to meet the sanitation goals of the UN MDGs. Movik and Mehta (2010) attribute much of the failure of previous efforts to 'top down, supply oriented approaches' that did not successfully engage local communities. CLTS contrasts with previous community-based initiatives as the emphasis is on 'facilitation' rather than simply focusing on 'educating and training' (Movik and Mehta 2010, p. 1).

Economic Benefit

In essence then, under the CLTS approach, socio-cultural and economic elements are integrated as complimentary and mutually reinforcing elements. Karmal Kar (the originator of the CLTS approach) and Katherine Pasteur (2005, p. iii) described CLTS as 'an approach which facilitates a process of empowering local communities to stop open defecation and to build and use latrines without the support of any external hardware subsidy'. At a broader economic level, a sanitation expert with the UN has asserted that on average, every US$1 spent on improving sanitation results in a US$9 economic benefit in the form of improved productivity and costs avoided due to improved health outcomes (UNHR 2011). For example, in a slum project in Tiruchirapalli, Tamil Nadu in India, the implementation of facilities for the safe disposal of waste, along with education in hand washing, resulted in a fall in household expenditure on medical treatments from 36 per cent to 4 per cent of household income (WaterAid and Tearfund 2002, p. 21).

One study focused on the economic returns of sanitation in the Cambodian context. The researchers calculated benefit–cost ratios of installed sanitation facilities in four rural locations in the Provinces of Battambang, Battambang Meanchey, Siem Reap and Kampong Thom (Heng et al. 2011). In each of the four sites, researchers used household questionnaires (1180 in total) along with other assessments and investigation methods. Along with the primary data obtained, national and local survey data was considered. Ultimately, benefit–cost ratios[3] were established for the sites for dry pit (simple), dry pit (solid), and wet pit latrines. The results are of course significantly more elaborate and conditional than is able to be outlined here. However, the broad results indicate that all latrine types returned a positive benefit–cost ratio such that the dry pits returned a benefit–cost ratio of 1.4 to 2 and wet pits returned a positive benefit–cost ratio of 2.3 to 2.9 (Heng et al. 2011).

Whilst the ratios may be instructive in regard to the economic benefit of the various latrine types, there are other issues to consider in terms of the practicality of affordability and usage at the village level. For example, the study found that 'unlined simple pit' latrines may only last around one year. However, their relatively low cost meant that poor households could afford them. In contrast, the cost of installing a more sophisticated 'wet pit' latrine at around US$170 makes the entry level more prohibitive for many. Complicating matters further though is that the annual cost of maintaining the higher cost 'wet pit' latrines was significantly lower than the annual maintenance cost of the initially cheaper 'dry pit' latrines (Heng et al. 2011). The researchers point out that programme costs per household should reduce with greater participation.

In addition, the researchers argue that properly implemented CLTS can lead to 'all households moving quickly onto the sanitation ladder' (Heng et al. 2011), a notion supported by earlier analysis by Karmal Kar in which he noted that 'Even where extremely poor families have had to return to open defecation temporarily in the event of latrine walls being eaten by termites or destroyed by heavy rain for example, they have tended to wait for the next productive season to earn money to purchase local materials for reconstruction' (Kar and Pasteur 2005, pp. 4–5).

The Market: Supply and Demand

An integral part of Kar's CLTS vision was the notion that solutions need to be market based, such that the local economy benefits. Kar (2005) has argued that villagers improving their sanitation access as part of the CLTS leads to the stimulation of the local economy as required hardware needs to be supplied. In 2004, Kamal Kar and Kazi Adil Ahmed Shafi undertook an assessment of the results of CLTS in several villages in Bangladesh where the CLTS approach had been previously implemented.

Kar and Shafi (Kar and Pasteur 2005, p. 3) note that where CLTS was appropriately facilitated, despite making certain observations of potential improvements 'most of the toilet slabs and pans were found to be clean and well maintained, and toilets were in use'. Furthermore, Kar and Shafi (in Kar and Pasteur 2005, p. 5) assert that the effect of 'CLTS has been to stimulate the entrepreneurial sector in producing and marketing latrine hardware, such as different types and grades of pans, rings and slabs. Local innovations have resulted in a range of differently priced products, and a spontaneous and competitive market has developed.'

In regard to the operation of the market in Cambodia, the Water and Sanitation Program (a partnership administered by the World Bank) conducted two market-based studies in 2006 and 2007. The first, in 2006, was a 'Sanitation Demand Assessment' and the 2007 study focused on 'Sanitation Supply Chain Analysis'. The 2006 'demand assessment' surveyed 939 households in 41 villages. Results of the survey indicated that whilst cost was not the only consideration, it was the major factor cited for not purchasing a latrine. Interestingly, many respondents who could not afford the relatively sophisticated latrine costing around US$150 still declared that they prefer to wait until they could afford such a system rather than spending less to purchase a less sophisticated interim latrine (Rosenbloom et al. 2011). This is an issue as only around 10 per cent of respondents said they could even afford a US$100 latrine. In contrast, around 50 per cent said they could afford a US$20 latrine (Rosenbloom et al. 2011). Clearly, the perception of the more sophisticated latrine (US$150) being the only purchase worth considering poses a major hurdle to getting people on the 'sanitation ladder' as discussed by Karmal Kar (2005) and others.

The 2007 study, based on interviews with 131 'actors' from the four main 'groups in the latrine supply chain' concluded that there was 'little coordination and limited information flow' in the supply chain. It was primarily constituted by 'small, independent enterprises' for which latrine components etc accounted for only a very small percentage of turnover (Roberts et al. 2007). In addition, masons, who may earn around 30 per cent income from latrine construction (Rosenbloom et al. 2011), were 'not particularly trusted by consumers, and are not seen as providers of reliable or credible information about sanitation' (Rosenbloom et al. 2011). Interestingly, only 60 per cent of masons in the study owned a latrine themselves as opposed to the other actors in the supply chain, leading the authors to conclude that 'as a group, masons may be less able than other actors to speak from experience when selling a latrine'(Roberts et al. 2007, p. 17).

The case of Tonle Sap

In light of the above discussion, it is worth giving some specific attention to Tonle Sap. The Tonle Sap basin covers over 44 per cent of Cambodia's area (ADB 2011). Tonle Sap Lake itself expands and contracts seasonally as water enters from the swollen Mekong River in the wet season and is dispelled into the Mekong in the dry season. Apart from the immediate health issues, Tonle Sap also accounts for around 75 per cent of Cambodia's fish catch (Flanagan and Considine 2011). The lake is home to many 'floating villages' where sanitation is inadequate with open defecation into the lake common.

The Chong Kneas area is located on Tonle Sap just south of Siem Reap. A RACHA (Reproductive and Child Health Alliance) survey was undertaken on seven villages in Chong Kneas (RACHA 2009). 88.6 per cent of the village households surveyed had no toilet or latrine and faecal matter is simply discharged directly into the river/lake (RACHA 2009). Innovative and sustainable approaches are being implemented though, with one example being the implementation of a 'bio-digester' sanitation system designed by Australian engineering students and engineers, in which methane gas produced is used by the community 'as a domestic energy source and the sanitised waste as fertiliser to improve crop yields on land around the lake' (Flanagan and Considine, 2011).

On a broader scale, the Asian Development Bank (ADB) Tonle Sap Rural Water Supply and Sanitation Project (TSRWSSP) took a market based approach (Robinson 2010). The goal of the TSRWSSP was to supply adequate water supply and sanitation facilities to around 1.09 million people in five provinces in the Tonle Sap Basin area (Robinson 2010, p. 21), which included building 150 100 latrines for households in the region by 2011. The success of the first phase of the TSRWSSP has led to the approval (by the ADB and the Cambodian Government) of a second phase of the project (Sethy 2012). Professor Sour Sethy of the University of Phnom Penh highlights the practical measures that have been taken to link the TSRWSSP to the national CMDG goals. For example, Sethy (2012) notes that a 'water quality index' was developed using data obtained from TSRWSSP fieldwork in participating provinces. 'Performance indicator' data sourced from 'government agencies, development partners', NGOs and relevant ADB reports are then utilized to assess the progress toward 'project and CMDG targets' (Sethy 2012).

The first phase of the TSRWSSP involved the installation of functioning latrines and generally required households to pay around US$137 toward the cost of materials, labour and installation, which is composed of: cash contribution = $90; in-kind materials = $46; labour = $1 (Robinson 2010). However, Robinson (2010, p. viii) asserts that data related to the project expenditure indicates that 'non-poor' households have utilized the project to build 'pour-flush latrines' as they are able to pay the required US$90 cash contribution, accounting for around 75 per cent of the project's 'sanitation investment'. In contrast, Robinson (2010, p. viii) says that the poor households were likely to have instead generally built cheaper 'dry latrines, so that only 'about 10 per cent of the sanitation investment appears to have reached those below the poverty line; the remaining 90 per cent has benefited non-poor rural households that took advantage of the generous subsidies offered by the ADB TS-RWSSP'.

Despite this clear disparity of access, proponents of the market-based models assert that economic input from recipients is essential. For example, the Canadian NGO, International Development Enterprises (IDE) (which is funded by the ADB among others) has also been active in Cambodia providing latrines at a cost to villagers, and IDE's sanitation programme manager has argued:

> When you give something to someone, there's a lack of appreciation . . . If you haven't mentally bought in to the concept of sanitation it's not likely that you're going to use it properly or maintain it properly. So the whole health benefit is moot. (Jacks in De Laney 2011)

CONCLUSION

In the case of Cambodia, it is clear that the challenges to be tackled in relation to sanitation are vast. Cambodia compares poorly to its neighbours and indeed globally on a national basis. There also exists an enormous disparity between urban and rural levels of access to improved sanitation. In tackling this urban and rural disparity, NGOs and other actors have been undertaking vital work within local village contexts. More specifically, the implementation of CLTS has been an important factor in translating goals to action and results. CLTS is an example of a framework for action that has been applied to the local context, something that clearly requires local social, cultural, economic and even political factors to be taken in to account. Such local level action is vital to addressing Cambodia's problems on a broader national basis and in regard to the urban/rural disparity.

In addition to such positive local action, there is also evidence of successful integration between international and national actors, such as the success of phase one of the Tonle Sap Rural Water Supply and Sanitation Project. What is perhaps less clear is a comprehensive integration between actors across all geopolitical levels. However, the achievement of UN and national goals related to sanitation access ultimately requires action within the local context. Local level actions, such as the implementation of CLTS, are essential if success is to be met at the national, regional or global level. Therefore, it may be timely to consider how that integration between geopolitical levels is best achieved and how this can translate to a coordinated approach at the local level. Whilst this chapter may fall short of offering a solution to the question of comprehensive integration among actors, making such a definitive assessment may be explored in subsequent research.

It is hoped that the overview of the sanitation issue in Cambodia has highlighted not only the scale of the problem, but also the potential for action afforded by some of the excellent work being undertaken by certain NGOs and agencies, particularly at local level. Therefore, despite Cambodia's relatively low levels of sanitation access, the potential for harnessing the efforts and energy of the various actors at all levels is cause for optimism. As the developing economies in Asia continue to grow, coordination of policy at all levels of government will be required in developing sanitation infrastructure and facilities. Sustainable sanitation infrastructure which caters to all citizens' needs and health outcomes is an essential part of the vision of a truly inclusive 'Asian century'.

NOTES

1. The authors propose a conceptual framework for public-private partnership that builds on a WHO framework for health delivery. See Figure 2 in Ejaz et al. (2011).
2. The CLTS foundation website has substantial information and good pictures pertaining to their work in Cambodia and other countries http://www.cltsfoundation.org/gallery_cambodia.htm.
3. The 'benefit–cost ratio' in this context refers to the ratio of the economic return per unit spent in currency terms (that is, generally in US$ benefit per US$ cost).

REFERENCES

ADB (Asian Development Bank) (2011), 'Tonle Sap rural water supply and sanitation sector', http://www.adb.org/projects/tonle_sap/basin.asp, accessed 16 January 2012.

CMP (Cambodian Ministry of Planning) (2010), 'Commune Database (CDB) and the implementation of Cambodian Millennium Development Goals (CMDG) at sub-national level', Phnom Penh, 6 December, http://www.un.org.kh/undp/media/files/CMDGprogress2007-2010_Eng.pdf.

Cambodian Ministry of Rural Development (2008), 'A toilet in every home: sanitation advocacy in Cambodia', http://www.wsp.org/wsp/sites/wsp.org/files/session3bsanitation.pdf, accessed 11 January 2012.

De Laney, Guy (2011), 'Mixing toilets with profit in Cambodia', 6 February, http://www.bbc.co.uk/news/business-12366108, accessed 20 December 2011.
Ejaz, E., B. Shaikh and N. Rizvi (2011), 'NGOs and government partnership for health systems strengthening: a qualitative study presenting viewpoints of government, NGOs and donors in Pakistan', *BMC Health Services Research*, **11**, p. 122, http://www.biomedcentral.com/1472-6963/11/122, accessed 5 June 2012.
Flanagan, Madeline and Mary-Lou Considine (2011), 'Australian engineering students "make it so" in Cambodia', *ECOS*, Issue 164, 10 October, CSIRO Publishing.
Heng, S., K. Poch, G. Hutton and P. Kov (2011), 'The economic returns of sanitation interventions in Cambodia', World Bank Water and Sanitation Program (WSP) Research Brief, August.
Kar, Kamal and Kath Pasteur (2005), 'Subsidy or self-respect? Community led total sanitation. An update on recent developments', IDS Working Paper 257, Institute of Development Studies: University of Sussex, November.
Movik, S. and L. Mehta (2010), 'The dynamics and sustainability of community-led total sanitation (CLTS): mapping challenges and pathways', STEPS Working Paper 37, Brighton: STEPS Centre.
RACHA (Reproductive and Child Health Alliance) (2009), 'Water and sanitation baseline survey: Chong Kneas, Siem Reap Province', Abstract, Phnom Penh.
Roberts, Michael, Aaron Tanner and Andrew McNaughton (2007), 'Supply chain assessment for sanitary latrines in rural and peri-urban areas of Cambodia', report prepared for The World Bank Water and Sanitation Program by International Development Enterprises, https://www.wsp.org/wsp/sites/wsp.org/files/userfiles/WSP-Supply-Chain-Assessment-Report-Cambodia.pdf, accessed 23 January 2012.
Robinson, Andy (2010), 'Sanitation, finance in rural Cambodia: review and recommendations', Asian Development Bank Water and Sanitation Program, May, p. viii., http://www.wsscc.org/sites/default/files/publications/wsp_adb_sanitation_finance_rural_cambodia_2010.pdf, accessed 4 February 2012.
Rosenbloom, Jan Willem et al. (2011), 'Sanitation marketing in Cambodia', *Waterlines*, **30** (1), January.
Sethy, Sour (2012), 'The Tonle Sap rural water supply and sanitation sector project of Cambodia', *Yale University Environmental Performance Index: Case Studies*, http://epi.yale.edu/epi2012/casestudies/reports/tonle-sap-rural-water-supply-and-sanitation-sector-project-cambodia, accessed 4 June 2012.
UN (2012), 'End poverty: 2015 millenium development goals', http://www.un.org/millenniumgoals/environ.shtml, 24 January.
UNESCAP (2011), 'Statistical yearbook for Asia and the Pacific 2011', http://www.unescap.org/stat/data/syb2011/I-People/Access-to-water-and-sanitation.asp, 5 January 2012.
UNHABITAT (2011), 'The Stockholm Statement to the 2012 United Nations Conference on Sustainable Development in Rio de Janeiro (Rio+20 Summit)', http://www.unhabitat.org/content.asp?cid=10244&catid=5&typeid=6&AllContent=1, accessed 12 December 2011.
UNHR (United Nations Human Rights), (2011) 'World toilet day, Saturday 19 November 2011 "Stop flushing your rights away, invest in sanitation," says UN expert on World Toilet Day', http://www.ohchr.org/en/NewsEvents/Pages/DisplayNews.aspx?NewsID=11616&LangID=E, accessed 25 January 2012.
WaterAid and Tearfund (2002), 'The human waste: a call for urgent action to

combat the millions of deaths caused by poor sanitation', http://www.wateraid.org/documents/plugin_documents/humanwaste.pdf, accessed 20 January 2011.

WHO/UNICEF (2010), Joint Monitoring Programme for Water Supply and Sanitation, 'Estimates for the use of improved sanitation facilities: Cambodia', updated March 2010.

Wicken, James (2010), 'Transforming lives – through water, sanitation and hygiene', Water Aid Australia, *Interaction*, December, **38** (4).

WSSCC (Water Supply and Sanitation Collaborative Council) (2012a), http://www.wsscc.org/resources/resource-statistics, accessed 4 January 2012.

WSSCC (2012b), 'More than 5000 children under five die every day from diarrhoea', http://www.wsscc.org/resources/resource-statistics/more-5000-children-under-five-die-every-day-diarrhoea, accessed 5 January 2012.

Zaidi, S., E.O. Hassan, S. Hodorogea, R.J.I. Leke, L. Tavara and M. Padilla de Gil (2010), 'International organizations and NGOs: an example of international collaboration to improve women's health by preventing unsafe abortion', *International Journal of Gynecology and Obstetrics*, **110**, supplement.

8. Decentralization and poverty reduction in Indonesia: the case of East Nusa Tenggara (NTT)

Yenny Tjoe

INTRODUCTION

Decentralization in Indonesia is principally implemented via two laws. Law 32/2004 on Regional Autonomy (original Law 22/1999) restructures the organizational arrangements of the regional government system, with a view to: (1) giving local governments at district and municipal level, greater representation, autonomy and resources; (2) giving emphasis to local knowledge and preferences about development; and (3) providing opportunities for local people to participate in decision making. The companion Law 33/2004 on Fiscal Balance between central and regional governments (original Law 25/1999) focuses on the intergovernmental fiscal system, intending to promote a more equitable distribution of resources, to increase responsiveness and fiscal capacity of local governments at district and municipal level, and to improve social welfare. By their stated objectives, these two laws depict the motives of Indonesia's 'Reformasi' (reform) in 1999, in order to reform a long lasting centralized autocratic regime into a public-participatory democratic system.

The outcome of these two key laws on decentralization, and more specifically on the alleviation of poverty, cannot be separated from two features. First, the role of central government transfers to sub-national levels to support the process of decentralization. And second, the commitment of local governments to allocate the resources for developmental purposes and poverty reduction. Although the latter plays a vital role in regards to empowering the poor, the former is equally important as a source of finance to support local governments that are lacking financial competency to perform regional autonomy. After a decade of reform, in terms of the role of central government, its fiscal transfer per capita has increased in value terms across all districts/cities and poorer districts are receiving larger amounts over time. However, the new sharing of revenues from taxes and natural resources sources has not led to the establishment of an effective means to redistribute resources from richer to poorer areas. Given the marked inter-regional disparities that affect the ability of various regions to raise local revenues and implement local projects, serious attention should be given to reformulating the allocation of transfers. On the other hand, in giving the widest possible fiscal power to 524 district/municipal governments, these two laws have created a diversity of local plans and objectives in which the pro-poor policies vary from region to region. In the case of poorer region such as Nusa Tenggara Timur (NTT) province, its district/municipal development expenditure per capita has transformed proportionally in favour of poorer districts over time. Such a shift in expenditure includes spending in strategic sectors of education, health, agriculture, forestry, livestock and fishery, food security, social affairs and cooperatives. Accordingly, out of the 20 districts and one city in NTT in 2010, 13 recorded a high rate of poverty reduction. This pro-poor local spending measure proves that decentralization in Indonesia has been working and has a positive impact in alleviating poverty.

This chapter is chiefly concerned with the effects of decentralization on poverty alleviation in Indonesia. The chapter looks at Laws 32/2004 and 33/2004 in the process of decentralization implementation in Indonesia, analyses the role of General Allocation of Funds and Shared Revenue as components of central transfers to regional governments, and finally examines the role of local policies in enhancing poverty reduction objectives through an NTT case study.

THE PROCESS OF DECENTRALIZATION IMPLEMENTATION IN INDONESIA

In May 1999, the People's Legislative Assembly of Republic of Indonesia (DPR) passed two laws to reorganize the intergovernmental system, namely Law 22 on Regional Autonomy and Law 25 on Fiscal Balance between central and regional governments. On 29 September 2004, based on the agreement achieved in *Rapat Paripurna* (the meeting of the People's Legislative Assembly), Law 22/1999 and Law 25/1999 were replaced by Law 32/2004 and Law 33/2004 respectively because the former were viewed as incompatible with the current development and demand of the regional autonomy programme. This amendment was needed because the first four years of reform (the first two stages of implementation process) were the period of formulating policy, working out strategy and implementation. The real strategy and formula began in 2004 (Figure 8.1).

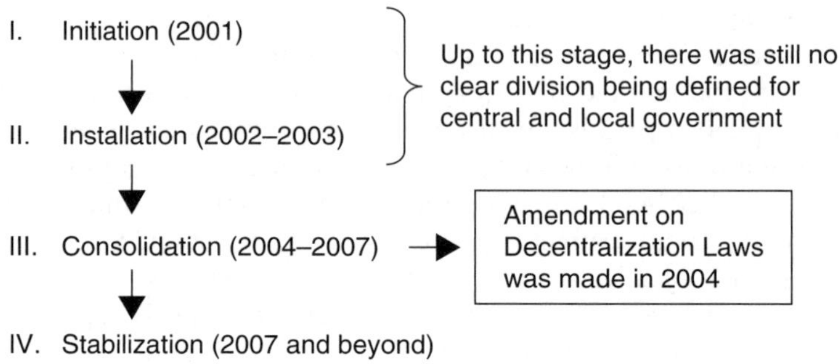

Source: Firman (2003).

Figure 8.1 Stages of implementation of decentralization process in Indonesia

Law 32/2004 on Regional Autonomy amended the original Law 22/1999 and restructures the organizational arrangements of the regional government system, including greater autonomy for local government at municipality/district level. Mayors or district heads act as local executives and prepare APBD (regional budget plan). They also consult with and report to locally elected DPRD[1] instead of the provincial governor. A local governmental organization called *Bappeda* (Regional Body for Planning and Development) is responsible for planning development strategies, public policy, and also directing the local financing structure. The *Bappeda* of each region can collaborate with foreign agencies or international non-governmental organizations (NGOs) for obtaining financial support or development assistance to enhance the process of regional development. However, it must be approved by a central government agency, that is, the international agencies or the NGOs need to register their proposals and obtain permits through the Ministry of Foreign Affairs (DEPLU).[2] The provincial governor acts as a coordinator and continues to report to the central government. Governors, mayors and district heads act as regional executives, and are given the self-government authority to manage their own affairs as appropriate. On the other hand, issues like defence and security, foreign policy, monetary and fiscal policy, judiciary affairs and religious affairs remain as national responsibilities (Table 8.1).

Law 33/2004 on Fiscal Balance between central and regional governments amended the Law 25/1999 and focuses on the intergovernmental fiscal system. The law includes more detailed clarifications about the role and duty of local and central governments, assigning a greater fiscal autonomy and responsibility to local governments at municipal or district

Table 8.1 Executive and legislative at national and sub-national levels

Level of government	Types of government	Executive	Legislative
National	Central	President	DPR (Council) (People's Legislative Assembly)
Sub-national	Province	Governor	DPRD (Provincial Local Council)
	Municipality	Mayor	DPRD (Municipal Local Council)
	District	Head of district	DPRD (District Local Council)

Sources: Law 32/2004 (UU No. 32, 2004) and Law 33/2004 (UU No. 33, 2004).

levels, in terms of taxing and expenditure powers. Mayors and DPRDs are responsible for the planning and implementation of the local budgets accordingly. The law also allows for borrowings made by local governments from both domestic and international sources with limited debt-to-revenue ratio,[3] which must be approved by both the DPRD and the Ministry of Finance, to maintain a healthy fiscal and monetary structure for the nation. The central government is committed to transferring a minimum of 26 per cent of its domestic revenues, called DAU (General Allocation Fund) to sub-national levels in supporting their process of decentralization. This transfer will be scrapped when local governments fail to perform their obligations.

Central Transfers: The Sharing of Revenues

Law 33/2004 allocates to each autonomous region *Dana Perimbangan* (Balancing Fund) to fund local needs during the implementation of decentralization. The balancing fund consists of three components:

1. Shared revenues derived from taxes and natural resources;
2. *Dana Alokasi Umum* (General Allocation Fund); and
3. *Dana Alokasi Khusus* (Special Allocation Fund).

Compared to the 'old sharing' arrangement in the pre-reform period, the Balancing Fund distributes more revenues from taxes and natural resources to the areas of origin. Regions that generate more revenues from taxes and natural resources receive larger share of revenues.

Consistent with the rhetoric claims of the two laws on decentralization, the new sharing arrangement gives local people (districts/municipalities) greater resources and opportunity in planning for their economic development. While the ultimate goal is to promote greater democracy and reinforce national unity, this appears more like a strategy to address political pressure from the separatist groups in resource-rich regions. Aceh and Papua, among others, are the major contributors of natural resource revenues but their administrations were monopolized by central government and their economy was increasingly dominated by the Javanese migrants, revealing the fundamental weakness of the centralized system. The conflicts have been gradually abated through peace negotiations after Aceh and Papua were granted special autonomy in 2001, giving these regions the right to protect and preserve local culture, ethnicity, heritage and religion.[4]

DAU (General Allocation Fund)

Article 27 of Law 33/2004 and article 37 of Government Regulation 55/2005 (PP No. 55, 2005) state that annual total DAU to all sub-national governments is maintained at a minimum of 26 per cent of national revenues. DAU, a sub-set of the Balancing Fund, is a form of subsidy to autonomous regions to achieve fiscal equality among regions. It replaces the grant system of the past, the SDO (Regional Autonomy Subsidy) and INPRES (Presidential Decree Fund), giving greater emphasis to equity and justice in the government transfer system. The proportion of DAU between a province and its districts/municipalities is set at 10 per cent and 90 per cent respectively, reflecting the task functions between government levels. The central government uses a formula to calculate the amount of DAU each district/city receives, taking 'local factors' into consideration (Figure 8.2).

Given the inter-regional disparities that affect the ability of various regions to raise local own-source revenue, DAU is distributed to aid local governments that are lacking financial capacity in their decentralization process. DAU of an individual region is calculated based on the sum of fiscal space and allocation base. Fiscal space refers to local capacity for spending on local needs with given local sources. Human Development Index (HDI) is one of the financing needs, indicating that regions with lower HDI will need more resources to improve the local quality of life, hence the pro-poor component of DAU. Allocation base refers to local government's needs, including their salaries, travel expenses, and other administrative expenses. As the central government is no longer responsible for local officials' salaries, the allocation base can be seen as assistance for local governments in their early phases of decentralization.

Fiscal space			+	Allocation base
Financing needs	–	Financing sources		
Factors:		Factors:		Factors:
• total population • area of region • GRDP per capita • HDI		• local own-source revenues • shared revenues from taxes and natural resources		• number of civil servants • their salaries and spending

Source: Author's interpretation based on Law 33/2004.

Figure 8.2 Formula to allocate DAU to a region

DAK (Specific Purpose Grants)

DAK is allocated to specific regions for financing infrastructure maintenance or enhancement; the purpose is aligned to national development priorities. Regions with low fiscal capacity and poor physical infrastructures are qualified to apply for DAK grants. Local governments that receive DAK are responsible for grant implementation and infrastructure utility evaluation based on the guidelines provided by the Ministry of Public Works.[5] Up to this stage the formula on the allocation of DAK is unclear and complex. The amount is decided by the Ministry of Finance with input from *Bappenas* (National Development Planning Agency), technical ministries and the Ministry of Home Affairs, based on the conditions of physical infrastructure and the region's capacity to finance such investment. When local governments fail to perform according to the guidelines, the Ministry of Finance will impose sanctions by withholding future DAK.

WELFARE OF VULNERABLE GROUPS IN THE POST-DECENTRALIZATION ERA

Decentralization is pursued as a new model of governance in Indonesia, and one of the stated objectives in the National Long-term Development

Plan (RPJPN 2005–25) is to enhance social welfare of its people in a faster way. The Indonesian term for social welfare is *kesejahteraan rakyat*. *Kesejahteraan* means happiness or prosperity, and *rakyat* means people, civilian, or grassroots. It is believed that the movement towards social welfare improvement in Indonesia has been related to unequal share of opportunity and wealth between central (Java island) and local (outer islands) and the subsequently increase in the gap between rich and poor. This is also relevant to the decentralization agenda of the post-Soeharto era, when the public demanded reform, and replaced the centralized governance with a more democratic system.

Hence, this chapter emphasizes the word *rakyat* and defines social welfare in the context of Indonesia. The welfare of society that requires improvement is the welfare of the vulnerable groups: the poor people who are most affected by the nation's unequal distributions of opportunity and wealth, and whose needs and concerns have been overlooked by the centralized system of public service. If Indonesia is to flourish and not to repeat the mistake of the former regime, poverty and inequality issues among regions must not be neglected. In particular, the welfare of vulnerable groups at the grassroots level becomes an important indicator to test whether today's Indonesia is utilizing its prosperity to reduce inequality between regions and to enrich the wellbeing of the poor.

In fact, the welfare of the grassroots level is one of the objectives in Indonesia's National Medium-Term Development Plan for the periods 2004–09 and 2010–14. The protection and empowerment of the rights of vulnerable groups (poor or minority) are key priorities of the new regime in the post-decentralization era. The increasing sign of peace in regions with ethnic or religious conflicts (Aceh, Papua, Maluku, and Central Sulawesi) indicates that the new government system has gradually improved the security and safety of the nation via democracy and decentralization of power to local levels. The plan also emphasizes the importance of pro-poor regional growth, which leads not only to creation of employment and better incomes, but also to improvement in access of the poor to clean drinking water and other public resources. Rural road systems are in particular a focus of the medium-term development spending to empower rural and vulnerable communities.

Another stated challenge is the commitment to achieve the Millennium Development Goals (MDGs) by 2015. In 2000, the Indonesian government made an ambitious goal to halve the proportion of its poor (from 1990 level) by 2015. Efforts to achieve this goal include increased national spending on non-energy subsidies from 2008–09 by 33.48 per cent (Kompas 2008) which includes giving away rice to poor households, and also supplying tools and seeds to farmers. In addition to this, the

Table 8.2 Share of pro-poor spending to total national expenditure (%), 2004–11

Year	Total DAU transfer to local levels / Total national expenditure (%)	Spending for education / Total national expenditure (%)	Spending for poverty reduction / Total national expenditure (%)
2004	30.23	n.a.	4.19
2005	28.87	n.a.	4.49
2006	31.59	n.a.	6.01
2007	33.79	11.80	7.44
2008	29.55	15.60	6.27
2009	30.83	20.81	6.61
2010	30.61	20.00	8.35
2011	31.23	20.21	5.99

Sources: Ministry of Finance (2012); Bappenas (2010); SAPA (2012).

government has increased its expenditure for development programmes by cutting subsidies on fuel and electricity since 2005 (World Bank 2007). This increased expenditure on development programmes has led to the increase in the share of national spending for public education, but direct spending for poverty reduction remains at less than 10 per cent of total national spending for the period 2004–11 (Table 8.2).

From an expenditure perspective, a small allocation of national resources for poverty reduction can mislead the local governments during their decentralization process. It may lead to a series of disoriented policies and budget objectives at the local level in which pro-poor policy may no longer be the priority. The commitment and leadership at the national level will need to be stronger to influence the local governments to give more weight to spending for the poor.

Pro-poorness of Central Transfers

From a poverty reduction point of view, the fairer distribution of natural resource wealth has not been pro-poor. The only revenue that is enjoyed equally by all districts and cities is fees and income from the fisheries sector. The winner in this new sharing arrangement is obvious: the regions that are rich in minerals, oil and natural gas. However, even in these resource-rich regions, the huge inflow of funds might not be appropriately utilized to address local problems and eliminate poverty issues. Many local governments were not ready for the sudden shift of power. Their policymakers

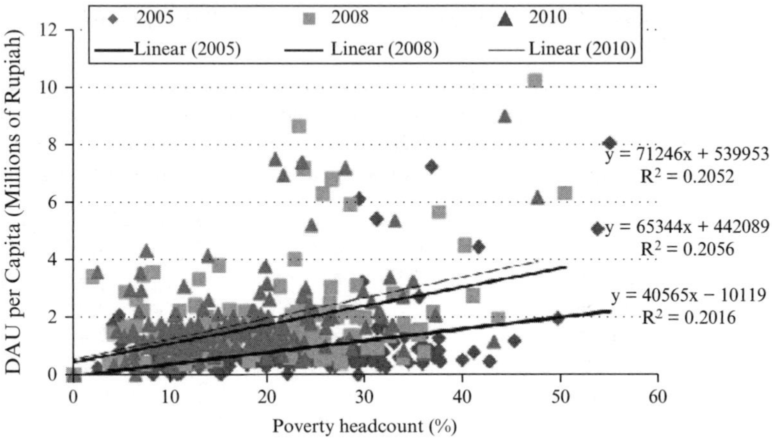

Sources: Presidential Regulation on DAU allocation and National Statistics Agency (2005, 2008 and 2007).

Figure 8.3 DAU per capita and poverty headcount, 2005, 2008, and 2010

and budget planners also failed to consult the local people about their plan, which resulted in unnecessary public spending. Brodjonegoro and Ford (2007) also warn that local governments in the resource-rich regions should be aware of the limited supply of natural resources and so the spending approach must be wise for a sustainable future.

In contrast to the shared revenues, the results of correlation analysis show that DAU allocation at the local level appears to be more pro-poor over time. In Figure 8.3, the trend line for 2010 is above 2008 and 2005, indicating the increasing amount of DAU received by local governments. The slope of line becomes steeper as the time horizon shifts, meaning that when the proportion of poor people in a district is high the DAU per capita it receives is also high. Overall, poorer districts/cities are receiving a greater amount of DAU per capita in the post-reform era but regions with fewer poor people also receive a greater amount of DAU. The pro-poorness of central transfers needs to be stronger in the future.

In the formula for DAU (Figure 8.2), fiscal space represents the capacity for financing local needs with local total revenues. A huge fiscal gap can exist in regions with a large proportion of poor people, no revenues from natural resources and under-developed infrastructure. Ideally, DAU transfers should reflect the fiscal gap proportionately to support local governments in their provision of public goods and services and to improve access of rural people to public places. However, the allocation of DAU also considers other factors such as the size of the region, population and the local government's needs, which minimizes the weight of pro-poor. The central government needs to give serious attention to reformulating the Allocation Fund objectives.

In terms of DAK, the grants have also increased over time but are mostly allocated for improvement of physical infrastructure in the education sector. Surprisingly, some regions in which the people rely on agriculture as their source of livelihood have received lower amounts of grants for irrigation and fishery sectors. Similarly, regions with poor access to clean water have received small amounts of DAK for the water sector. The allocation of DAK funds to a region appears to be heavily based on technical measurement rather than the proportion of infrastructure in supporting and enhancing social welfare. Another issue is local governance in managing DAK. When proposing DAK grants, local governments and local communities are involved in a development plan, meeting to decide which areas need improvement. However, participants have complained that the whole process of meeting and planning lacked transparency. After implementing the plan, the governments showed lack of accountability in monitoring and evaluating the public facilities that had been built by using the grants (Usman et al. 2008).

A study conducted in 2004 (Suhendra and Amir 2006), about the impact of fiscal decentralization on local governments' financial ability, found that the majority of Indonesian local governments lack financial capacity. They have not had the ability to improve own source revenues (PAD), and have a very low share in total local expenditure. Out of 440 districts/ municipalities in 2004, 214 were observed. Only 19 had high growth of PAD with a higher share in total spending. The intention of distributing DAU to promote equity and justice as well as DAK to enhance public infrastructure seems to be taken for granted by local governments while their accountability to local people has received many criticisms.

Thus, concerns have arisen from this overreliance on central transfers that either local officials are misusing the decentralized autonomy for their own interests or they are failing to perform their duties. Many local governments repeatedly propose a high local expenditure in their APBD (annual local budget) while in practice most spending is allocated for routine expenses, which has been criticized as local officials' attempt to ensure that more transfers are continually received (Brodjonegoro 2003; Sjafrizal 2001). Furthermore, when local governments fail to allocate their expenditures according to their proposed APBD, they will have either their DAU slashed or DAK withheld for the next term. Using expenditure as a measurement of local government performance can potentially encourage misuse of funds. Thus corrupt practice has existed since the previous regime, and there is concern as to whether corruption has simply been transferred to local governments in today's decentralized Indonesia.[6]

DECENTRALIZATION FOR POVERTY REDUCTION: PRO-POOR LOCAL POLICIES

Decentralization is believed to offer advantages in reducing poverty and inequality. A decentralized public expenditure can yield better quality of public works (such as sewers, roads and water) or larger quantities of services at a lower cost because it avoids high administrative costs implied by centralized systems (Martinez-Vazquez and Mcnab 2003). The direct contact with local people can provide advantages to local policymakers in providing suitable services to the low-income earners, while reducing the transaction costs in identifying the poor (Oates 1999). Decentralization will also encourage initiatives of local governments to deregulate economic policies in facilitating development (Rondinelli 1990), leading to creation of jobs and improvement in local income. Nevertheless, effectiveness of decentralization in eliminating inequality and poverty will depend greatly on the pro-poorness of local policies in individual regions.

Policy is pro-poor when the outcomes favour the poor, in that poverty can be reduced at a faster rate than otherwise. Pro-poor policies include public participation and community planning that empower the poor (Chavis 2009; Justino 2006; Olson 1971), and use of local public spending that reflects a more pro-poor manner (Mahal 2000), including transfers to the target groups, rural infrastructure investment, agricultural developments, and low-interest credits for small-scale industries (Braun and Grote 2000). Pro-poor local policies require a willingness of local governments to consult the local community, including the poor, in their policy and budget planning. This is also a way to ensure that decentralization does not become a paradise for local officials collaborating with local elites at the expense of the welfare of poor people. As pro-poor public policies at the local level matter, the role of local governments and local councils becomes very crucial in the whole process of decentralization (Hossain 2003). Their policy designs must ameliorate inequality, so that local economic growth can lead to poverty reduction (Islam 2003). To achieve poverty reduction and reduce inequality, pro-poor public spending at local levels matters.

Local policies in Indonesia are reflected in their planning and budgeting. Given the marked inter-regional disparities in financial resources, it is essential for local governments to carefully allocate resources in order to maximize public benefits. The Ministry of Home Affairs reported that for the period 2006–11, almost half of the local resources in 297 district/municipal levels were allocated to routine expenditure which includes local government's salaries, travel and other expenses. The number increased from 176 districts/cities in 2008. This leaves only a tiny share of expenditure for key sectors such as education, health, and infrastructure. It also poses a question about the overall pro-poorness of local policy objectives, as to whether the outcomes favour the low income groups (farmers, fishermen, and informal workers) and those in rural (remote) areas which lack infrastructure, to allow access to health centres, schools, and clean water.

Pro-poorness of Local Spending in Indonesia

In giving the widest possible fiscal power to all the district/municipal governments, Law 32/2004 and Law 33/2004 have created a diversity of local plans and objectives. The level of pro-poorness of local policies varies from region to region. This can be analysed from the total expenditures in 2010 published by the Ministry of Home Affairs. The results show that in aggregate, the relationship between total spending per capita and poverty headcount in districts and cities in Indonesia is positive but weak ($r = 0.30$). Among the four poorest provinces of Indonesia, West Papua, Maluku show a positively strong correlation, while Papua shows no relation. West Sulawesi, the fourth poorest, is the only region in which its districts' spending pattern shows a strong negative relationship to the proportion of poor people. Out of 32 provinces being tested, the spending patterns of districts in 17 provinces show either weak or no relation to poverty rate. This implies that the majority of local policies for 2010 are lacking emphasis on the number of poor people in each region.

To measure the pro-poor outcome of this set of local policies for 2010, the rate of change of provincial poverty between 2010 and 2011 was observed. All provinces experienced a negative rate of growth in poverty headcount, except for South Kalimantan and West Sulawesi; both recorded a 2 per cent growth per annum in the proportion of poor people for the period 2010–11. The highest rate of decline, 17 per cent per annum, occurred in Maluku. East Kalimantan recorded an annual declining rate of 12 per cent and West Papua 8 per cent; both are above the national average (6 per cent decline per year). Although there is no conclusion on the relation between total local spending in 2010 and poverty reduction for the period 2010–11, a strong inverse relationship between local spending and proportion of poor people may harm those at the lower quintiles of society, especially in the poorer regions.

The pro-poorness of local spending can also be analysed from the perspective of the DAU formula. Although the local governments' needs is one of the factors (allocation base) used in the calculation of DAU, the funds are transferred by central government to each autonomous region as assistance in the provision of public goods and services. The local governments should carefully allocate their resources to ensure they meet their responsibility to the central government as well as the local people. A high share of routine expenditure is tolerable especially in the early phases after decentralization is implemented, given the increased responsibilities as well as the number of government personnel and public servants at local levels. However, allocating half of the government resources to finance the government administrative needs hardly reflects a pro-poor policy. Hence, it is not surprising to see a weak correlation between total local spending and poverty rate. In the stabilization phase (the last stage of decentralization) this trend of large routine spending on government administration is only going to build up popular cynicism in which local civil society criticizes local pro-poor policies and perceived corrupt officials in the local governments.

PROGRESS OF LESS DEVELOPED REGIONS: CASE STUDY OF NTT

NTT, grouped into an isolated region (Hill and Weidermann 1989), is the fifth poorest province in Indonesia (Bappenas 2010). The province comprises 21 districts and one municipality. Despite the economic boom during the Suharto regime, this region has never enjoyed the prosperity of the country. In 1990, nearly 40 per cent of villages in this province were living under the poverty line (BPS 1992). Just before Indonesia began its reform and decentralization programme, NTT was still very poor and was the leading province in terms of poverty incidence and infant malnutrition (Hugo 2004). It is one of the regions that urgently need improvement, especially in the sectors of education, health, nutrition, and sanitation facilities.

Since implementation of reforms, NTT's economy has developed steadily. Agriculture, the main source of livelihood for most of the population in NTT, contributed nearly 41 per cent of NTT's gross regional domestic product (GRDP) in 2010, and employed the majority of the working population (71.27 per cent in 2010, down from 79 per cent in 1994). Poverty in NTT is largely due to the dependence on farming of much of the population, and an under-developed road system which cuts off the access to markets and public places. In some rural areas, it is still common to find subsistence farming activities where people grow corn and other basic foods for their own consumption rather than for income. This province is at high risk of food insecurity when disasters happen.

Pro-poor Outcomes of Decentralization in NTT

First is the establishment of a special team to address poverty issues at local levels. In 2002, the NTT provincial government established the NTT Provincial Committee to Combat Poverty (*Komite Penanggulangan Kemiskinan Pemerintah Provinsi NTT*).[7] More recently, alleviation of poverty became a nationwide programme. As stated in Presidential Decree No. 54/2005, the Coordination Team to Combat Poverty (TKPK – *Tim Koordinasi Penanggulangan Kemiskinan*) functions to reduce the poverty in Indonesia. Accordingly, the NTT Provincial TKPK has accomplished a number of tasks (NTT Provincial Government 2006; NTT Provincial Bappeda 2007), as follows:

- In 2006, established one TKPK for each district and municipality.
- Developed strategies to combat poverty at provincial level period 2004–10; currently working on strategies for district/municipal levels.
- Increased cooperation between local governments and NTT Central Bureau of Statistics.
- Created a single database on poor households in the NTT region (monetary and non-monetary characteristics of households at each sub-district and village), annually updated.
- Trained 30 staff (NGOs and public servants) in the methods of identifying the poor, and continually developing the 'Participatory Poverty Assessment'.[8]

Second, there has been increasing assistance from domestic and international NGOs that aims to enhance the welfare of the poor. Since 2004 development projects have begun in NTT with various objectives, among others to encourage community awareness of the importance of education for poor children, provide scholarship assistance, improve access to clean water, improve the health and nutrition of rural young children, improve resilience of people through disaster management and risk reduction training, and promote small-farm business development. Most projects are in collaboration with *Menkokesra* (Coordinating Ministry of People's Welfare) and the implementations are carried out by selected local NGOs which possess intuitive local knowledge and familiarity to local villagers. Local communities are involved in the decision-making process.

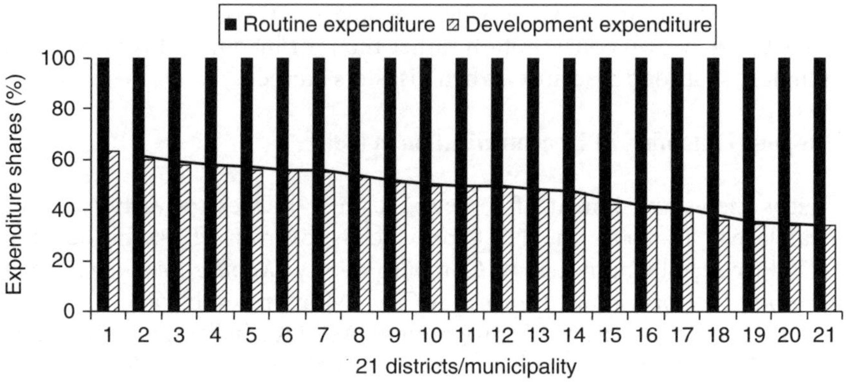

Sources: NTT-PEA (2009); Ministry of Home Affairs (2012) on APBD Report.

Figure 8.4 Development and routine expenditures at districts/city in NTT (%), 2010

To enforce the participation and commitment of local people, they are required to contribute to the project's cost, either in cash or in labour.

Third, pressure from civil society and central government in terms of the realization of local spending has forced local governments and policy-makers to be more accountable in allocating their resources. The share of development expenditures in NTT have gradually increased since 2009. Out of 21 local governments in 2010 and 2011, 10 districts and 1 city are consistently allocated less than 50 per cent of total spending for routine expenditures (local government salaries and other spending) (Figure 8.4 and 8.5). Figure 8.5 shows that the proportion of development spending in all 21 local governments has become higher (slightly) compared to the previous year. Ten districts also showed declining share for government needs. Eliminating poverty may take more time than expected, but with less routine expenditure, more resources are available for community or infrastructure development.

The NTT province has experienced improvement in terms of human development and quality of life. Its proportion of poor has fallen from 36 per cent in 1999 to 23 per cent in 2010. The proportion of people living under the poverty line declined sharply in 17 districts and one city between 2008 and 2010. The improvement in HDI, and the decline in poverty rate, are positive signs of decentralization. However, several areas require serious attention and immediate action: in particular access for the vulnerable groups to public resources such as public schools, health clinics and clean drinking water. Many children and pregnant mothers cannot

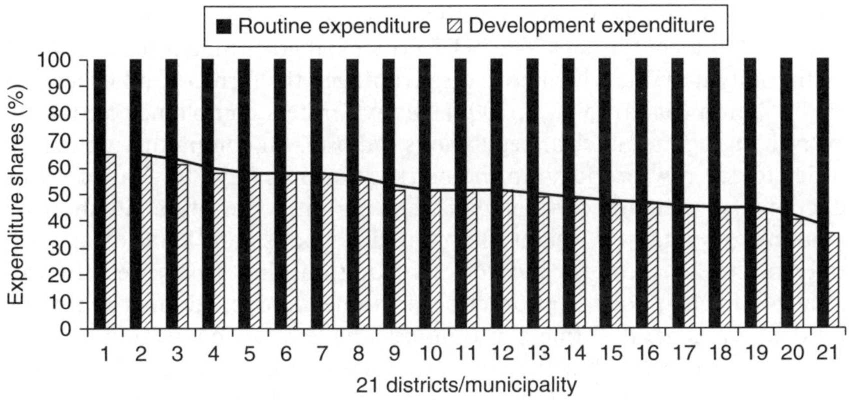

Sources: NTT-PEA (2009) and Ministry of Home Affairs (2012) on APBD Report.

Figure 8.5 Development and routine expenditures at districts/city in NTT (%), 2011

afford to have regular health check-ups, eat sufficient nutritious foods, or drink clean water, resulting in high rates of child malnutrition and infant mortality. Given the fluctuation in food prices and limited access to health services, many young children in rural NTT continue to have insufficient calories and nutritional intake.

Criticism of Local Policies

Despite facing the arid conditions and natural disasters in the province, the local government in NTT is reported to take no concrete actions to cope with the food and clean water crisis. It is common to hear local people criticizing their local governments as being not responsive to local complaints and input. The local development plan has established food security committees to secure village food supplies. Health workers, midwives, village leaders and religious leaders are involved as the source of ideas and input. However, only a few villages perform effectively in this programme, and the government did not engage the progress further in other villages. Similarly, the government in one of the districts provided water pipes in local villages, but for years it failed to provide the water. After a series of complaints with no response from the government, the villagers decided to fix the pipe to connect to a water source, costing each household Rp2000 (US$0.20). Despite proposals on village development submitted by village leaders, government development policies rarely implement these ideas.

At the grassroots level in the farming districts, local policies have been criticized as being inconsistent and anti-poor as they have tended to neglect the needs of local farmers. The NTT provincial government regulate local commodity markets which have disadvantaged the farmers and fishermen in the region (Suharyo et al. 2007). Many farmers complained about the ever-changing and unclear regulations imposed on commodity goods. In order to sell raw products to the market, farmers have to show a set of documents, including certificate of trade permit, letter of details showing amount of quota and weight of commodity goods, and letter of quarantine and health check for livestock. In addition, there is a series of levies attached to each document or letter mentioned above, adding to farmers' burdens. In general, farmers in Indonesia have rarely completed primary education, meaning that they need to obtain permits through agents (or specialists in administrative documentation), which often costs much more than the original fee when completed and submitted by themselves. Meanwhile, NTT farmers who work small landholdings (less than 0.5 ha) have small harvests, resulting in weak bargaining power. Middlemen and big merchants, on the contrary, possess considerable power, as they buy from farmers at lower prices while selling at higher prices to the market.

Additionally, in urban areas such as in Kupang city, pro-poorness of local policy is also questioned. Only a small percentage of total expenditure is allocated for helping the poor (Therik 2010; Nedabang and Sape 2008), through maternal and child healthcare, food security programmes via farmers' empowerment, and empowerment of poor villages to overcome the impact of crises. However, after being ranked as the most corrupt city in 2008 by Transparency International Indonesia,[9] the commitment of local authorities to meet the needs of local people has become stronger. The city's Corruption Perception Index has improved from 2.97 in 2008 to 4.89 in 2010 (TI-Indonesia 2010). Local residents have experienced the benefits of reform, and the city's administration officials have heightened accountability to eradicate corruption.

CONCLUDING REMARKS

There are several points arising from Indonesia's decentralization on the overall efforts to reduce poverty in the country. The first is the real delegation of power to lower tiers of government (district/municipality) rather than merely de-concentration. This has given poor and remote regions a chance to gain a fairer share of resources. Districts and cities with a high proportion of the poor have continuously received greater amounts of central government transfers via DAU. However, under-resourced regions in Indonesia still lack adequate influence in national politics due to their small contribution to national revenues. As these regions have small local own-sourced revenues, they have to compete with other regions in obtaining resources from central government. The decision on the amount of DAU cannot be based on how much is needed; instead it should be based on what areas need funding. Similarly, the decision on DAK allocation should put more emphasis on the quality and utility of investment. There is also a problem with the DAU cut off and DAK withholding when local governments fail to carry out expenditures according to their proposed APBD. This measurement of performance is not contributing to poverty reduction; on the contrary it has facilitated many large-scale development projects that are unnecessary or not effective for supporting the livelihood of local people. Evaluation of local governments should put more emphasis on their performance in human development.

The second point concerns the commitment of the new governance system in its promise to create a safe and peaceful environment for every citizen. Although the special autonomy regions granted for Aceh and Papua can be seen as preventing further separatist movements in conflict areas, decentralization reform is not merely for reinforcing national unity. The desire to improve social and economic conditions of the outer islands and the vulnerable groups exists and is clearly stated in Law 32/2004 and Law 33/2004, as well as in the objectives of the MDGs and the National Long-Term Development Plan (2005–25).The new system has given greater protection of human rights, freedom to preserve local heritage and respect to local preferences. The movement towards welfare enhancement has indeed attracted numerous aid agencies into Indonesia to carry out humanitarian activities. Under-developed regions that urgently need improvement in education, health, and livelihood have become the destinations of development projects and ongoing funds are flowing into these regions. The case study of NTT province shows that poverty reduction programmes have been localized to district/municipal levels. Overall human development improves and more households can afford spending well above the poverty line. To relieve poverty, development initiatives in the future will need to directly target the health of women and children in the rural and remote regions.

The third point is the emergence of civil society and freedom of the press in post-reform Indonesia. These two actors were heavily suppressed during Soeharto's presidency. There is an important role that civil society and the media play in poverty reduction and protecting the welfare of the vulnerable. They help to exert pressure on governments and to control government actions whenever injustice happens. Students, local NGOs, local newspapers, and other activist groups have become very vocal in recent years, and there have been forthright protests against the government's lack of responsiveness in handling issues such as fuel price increases, burning of religious facilities, and administrative scandals. In facing dissatisfaction and criticism from society, Kupang city local authorities were forced to develop a more responsible model of governance. A decreased level of corruption is a sign of a more responsible allocation of resources. The paradigm of unbalanced spending between local governments' salaries and spending for the public is still strong in local governments. This dominant spending for local governments' needs in the early phases of reform will eventually need to be reduced. Unless a credible reason is communicated to the public, any increase in routine spending in future would possibly lead to loss of popularity for district leaders or mayors, which will often ensure their defeat in the next election.

In sum, decentralization in Indonesia does enhance poverty reduction and also an increased level of pro-poorness in the nation. This pro-poor decentralization is important to ensure sustained poverty reduction in the future. What the new system needs to enforce is transparency, participation and policy coherence at the local level in order to eliminate the inefficiencies. There should be regular evaluation to measure these three qualities of local governments. The results should be published and the best leaders rewarded as an incentive for others to emulate and be more accountable to their people.

NOTES

1. DPRD stands for *Dewan Perwakilan Rakyat Daerah*, meaning Regional People's Legislative Assembly. It functions as a local legislative body and has the right to audit local budge. DPRD is equivalent to local council under the Australian local government system.
2. DEPLU stands for *Departemen Luar Negeri*, or Ministry of Foreign Affairs. Laws on International agencies or NGOs include Law No. 37/1999 on guidelines for registration of international NGOs and Law No. 11/2009 which states that central government develops networks and coordination for international NGOs in the implementation of social programmes.
3. Total debt is limited to 75 per cent of revenue minus necessary expenditure. This condition was allowed but not stated directly in Law 25/1999.
4. See Brodjonegoro and Ford (2007) for more comprehensive discussion on special autonomy for Aceh and Papua.
5. The Ministry of Public Works provides Ministry Regulation No. 42/PRT/M/2007 on the implementation, monitoring and evaluation of the DAK for infrastructure sectors: irrigation, sanitation, drinking water, and road works.
6. Transparency International places Indonesia in 100th position out of 183 countries. The survey is based on CPI (Corruption Perception Index), with a scale of 0 to 10, 10 being a highly 'clean' country. Indonesia scores 3.0 in 2011, improved from 1.7 in 2001.
7. In 2007, the name of this committee was changed to the NTT Provincial Coordination Team to Combat Poverty (*Tim Koordinasi Penanggulangan Kemiskinan Provinsi NTT –*

TKPK Prov. NTT). The functions of this team are stated directly in NTT Governor Decree (KepGub) No. 44/KEP/HK/2007.

8. Participatory Poverty Assessment is a research process that includes poor people's views in the analysis of poverty and involves them directly in formulation of strategies and follow-up action to reduce poverty.
9. The survey gathered information from respondents in 50 cities across Indonesia (33 provincial capitals and 17 cities with good economic performance), asking how local businesses and citizens perceived their local governments, and whether corruption was common in their society. On the CPI (see note 6), the most corrupt city in 2010 is Pekanbaru (CPI score: 3.61).

REFERENCES

Bappenas (2010), *Achieving The Millennium Development Goals in Indonesia (Laporan Pencapaian Tujuan Milenium di Indonesia)*, Jakarta: Ministry of National Development Planning Agency.

BPS (1992), *Statistik Kesejahteraan Rumahtangga 1991* (National Economic Survey), Jakarta: Biro Pusat Statistik.

Braun, J.V. and U. Grote (2000), *Does Fiscal Decentralization Serve the Poor?*, IMF Conference on Fiscal Decentralization, Center for Development Research (ZEF-Bonn), Washington, DC.

Brodjonegoro, B. (2003), *Decentralization and Business*, The Indonesian Update.

Brodjonegoro, B. and J. Ford (2007), 'Inter-governmental fiscal relations and state building: the case of Indonesia', in R.M. Bird and R.D. Ebel (eds), *Fiscal Fragmentation in Decentralized Countries: Subsidiarity, Solidarity and Asymmetry*, Cheltenham, UK and Northampton, MA, USA: Edward Elgar, pp. 348–53.

Chavis, L. (2009), 'Decentralizing development: allocating public goods via competition', *Journal of Development Economics*, doi:10.1016/j.jdeveco.2009.10.001.

Firman, T. (2003), 'Potential impacts of Indonesia's fiscal decentralization reform on urban and regional development: towards a new pattern of spatial disparities', *Space and Polity*, **7** (3), 247–71.

Hill, H. and A. Weidermann (1989), 'Regional development in Indonesia: patterns and issues', in H. Hill (ed.), *Unity and Diversity: Regional Economic Development in Indonesia Since 1970*, Singapore: Oxford University Press, pp. 3–54.

Hossain, M. (2003), 'Decentralization and development: what prospects in Bangladesh?', in M. Hossain, A.N. Nabi and I. Islam (eds), *Bangladesh's Development Agenda and Vision 2020: Rhetoric or Reality?*, Dhaka: The University Press Limited, pp. 289–300.

Hugo, G. (2004), 'International labour migration and rural dynamics: a study of Flores, East Nusa Tenggara', in T.R. Leinbach (ed.), *The Indonesian Rural Economy: Mobility, Work and Enterprise*, Singapore: Institute of Southeast Asian Studies (ISEAS), pp. 103–33.

Islam, I. (2003), 'Human development, democracy and economic growth: Bangladesh in the 21st century', in M. Hossain, A.N. Nabi and I. Islam (eds), *Bangladesh's Development Agenda and Vision 2020: Rhetoric or Reality?*, Dhaka: The University Press Limited, pp. 39–55.

Justino, P. (2006), 'The impact of collective action on economic development: empirical evidence from Kerala, India', *World Development*, **34** (7), 1254–70.

Kompas (2008), *Subsidi Naik Rp15,84 Triliun, Hindari Penyaluran yang Salah Sasaran dengan Fokus yang Jelas*, Jakarta, Indonesia: Kompas, 31 October.
Mahal, A.S. (2000), 'Decentralization and public sector delivery', Discussion Papers on Development Policy, Bonn: ZEF.
Martinez-Vazquez, J. and R.M. Mcnab (2003), 'Fiscal decentralization and economic growth', *World Development*, **31** (9), 1597–616.
Ministry of Finance (2012), 'Directorate General of Fiscal Balance', Ministry of Finance, Republic of Indonesia, http://www.djpk.depkeu.go.id accessed 20 March 2012.
Ministry of Home Affairs (2012), *Directorate General of Regional Finance*, Ministry of Home Affairs, Republic of Indonesia, http://www.djkd.depdagri.go.id, accessed 25 May 2012.
Nedabang, A. and A. Sape (2008), *Laporan Utama Pos Kupang Februari 2008: Mencermati Anggaran Pembangunan NTT 2007* (*The Top News in February 2008: Observing the NTT Provincial Development Expenditures in 2007*), Pos Kupang (Kupang City).
NTT-PEA (2009), *East Nusa Tenggara Public Expenditure Analysis*, Kupang: NTT Research Team.
NTT Provincial Bappeda (2007), *Rekor Penanggulangan Kemiskinan* (*Combating Poverty Record*), Bappeda Prov. NTT, http://www.bappeda.nttprov.go.id, retrieved 3 January 2011.
NTT Provincial Government (2006), *Poverty Dimension and Poverty Reduction Policy in NTT Province*, Kupang City, Nusa Tenggara Timur, Indonesia. Retrieved 3 January 2012 from Kantor Pengolahan Data Elektronik (electronic data processing office).
Oates, W. (1999), 'An essay on fiscal federalism', *Journal of Economic Literature*, **37** (3), 1120–49.
Olson, M. (1971), *The Logic of Collective Action: Public Goods and the Theory of Groups*, Cambridge, MA: Harvard University Press.
PP No. 55 (2005), *Peraturan Pemerintah Republik Indonesia Tentang Dana Perimbangan* (*Government Regulation on Balancing Fund*).
Rondinelli, D. (1990), *Decentralizing Urban Development Programs: A Framework for Analyzing Policy*, Washington, DC: US Agency for International Development.
SAPA (Strategic Alliance for Poverty Alleviation) (2012), 'Data and information on poverty in Indonesia', Strategic Alliance for Poverty Alleviation, http://www.sapa.or.id, accessed 6 March 2012.
Sjafrizal (2001), *Regional Autonomy Impact: The Case in West-Sumatra*, Jakarta.
Suharyo, W., Toyamah, N., Poesoro, A., Sulaksono, B., Usman, S. and V. Febriany (2007), *Iklim Usaha di Provinsi NTT: Kasus Perdagangan Hasil Pertanian di Timor Barat*, Jakarta: Lembaga Penelitian SMERU (Research Institute).
Suhendra, M. and H. Amir (2006), *Fiscal Decentralization in Indonesia: Current Status and Future Challenges*, Republic of Indonesia: Ministry of Finance.
Therik, W. (2010), *Hasil Analisis APBD Kota Kupang T.A. 2010* (*Results of analysis on Kupang City's Budget Plan Financial Year 2010*), an analysis for discussion in AMPEK (The Kupang City's Grassroots Alliance).
TI-Indonesia (2010), *Measuring Corruption in Indonesia: Indonesia Corruption Perception Index 2010 and Bribery Index*, Jakarta: Transparency International Indonesia.
Usman, S., S. Mawardi, A. Poesoro, A. Suryahadi and C. Sampford (2008), *The*

Specific Allocation Fund (DAK): Mechanisms and Uses, Jakarta: The SMERU Research Institute.

UU No. 32 (2004), *Pemerintahan Daerah* (*Law on Regional Autonomy*), amendment of UU No. 22/1999.

UU No. 33 (2004), *Perimbangan Keuangan antara Pemerintah Pusat and Pemerintah Daerah* (*Law on Fiscal Balance between Central and Regional Governments*), amendment of UU No. 25/1999.

World Bank (2007), *Spending for Development: Making the Most of Indonesia's New Opportunities*, Washington, DC: World Bank.

9. Migration of Bangladeshi workers to Malaysia: emerging lessons of economic and social costs and benefits at the migrant, migrant household and community levels

Munshi Israil Hossain, M. Adil Khan and Patricia Short

INTRODUCTION

Asia's varying economic performance, changing demographic characteristics and improved transportation, stimulated further by social and economic inequalities, both within and across nations, and expanding labour demands in the faster growing economies of the region, have triggered the emerging trend of Asia to Asia migration (A2AM) in the Asian region (UN, 2004; Hugo, 2005).[1]

The A2AM has been bolstered by three successive waves of rapid economic growth in some parts of Asia: Japan led the first wave; South Korea, Taiwan, Hong Kong and Singapore (the newly industrialized economies–NIEs) the second wave, and Malaysia, Thailand and Indonesia the third wave. These growth waves that accompanied expanded economic activities against the backdrop of falling rates of population growth and shifts in the labour market have produced labour shortages, especially in the semi and unskilled sectors, creating opportunities for migration of workers. In the emerging scenario of A2AM Bangladesh, a least developed country of Asia, has become a major labour sending country, and Malaysia a major receiving country of the Bangladeshi workers. This chapter discusses the political economic dimension – the 'pull' and 'push' factors – of migration of Bangladeshi workers to Malaysia.

MIGRATION OF BANGLADESHI WORKERS: THE BACKDROP

In recent years, the economy of Bangladesh has experienced significant structural transformations – from a predominantly agriculture-based economy. The share of manufacturing and services has increased significantly and thus has created, albeit at a rate slower than that of population growth, new job opportunities, mostly in the urban areas.

Also in recent times, while Bangladesh's population has increased from 81 million in 1980 to 149 million in 2010 (World Bank, 2011), the labour force participation rate has fallen from 75 million in 1990 to 71 million in 2011 (CIA, 2011) and during this period, the unemployment rate has risen from 1.8 per cent in 1985 to 4.8 per cent in 2010 (CIA, 2011).[2] Furthermore, between 1980 and 2010, the value of the Gini coefficient, a measure of inequality (the higher the value, the more unequal the income distribution) rose from 0.25 to 0.38 (World Bank, 2011) signalling that Bangladesh's growth has been somewhat unequal and less elastic in terms creation of jobs vis-à-vis demand.

During the same period remittances from overseas workers increased from USD301.33 million in 1980 to USD 11 004.73 in 2010, meaning that migration and remittances have crucial importance in the economy of Bangladesh, so much so that migration and remittances are estimated to have reduced poverty by about 6 per cent in the country (World Bank, 2007).

Since 1980 an estimated 7.25 million Bangladeshis have migrated overseas on short term employment contracts (BMET, 2011). Bangladesh is now ranked seventh among the manpower exporting countries of the world (World Bank, 2011). More than half of all Bangladeshi migrants are unskilled, followed by skilled (31.3 per cent), semi-skilled (14.8 per cent) and professionals (2.4 per cent).[3] Over this period and up to 2011, a total of 700 397 Bangladeshi workers have migrated to Malaysia (BMET, 2011). Figure 9.1 illustrates the trend of migration of Bangladeshi workers globally.

MIGRATION OF BANGLADESHI WORKERS TO MALAYSIA: THE POLITICAL ECONOMIC DIMENSION

The migration of Bangladeshi workers to Malaysia is a recent phenomenon and although the flow of migrants has been somewhat uneven, an overall upward trend has continued, with noticeable increase in the last

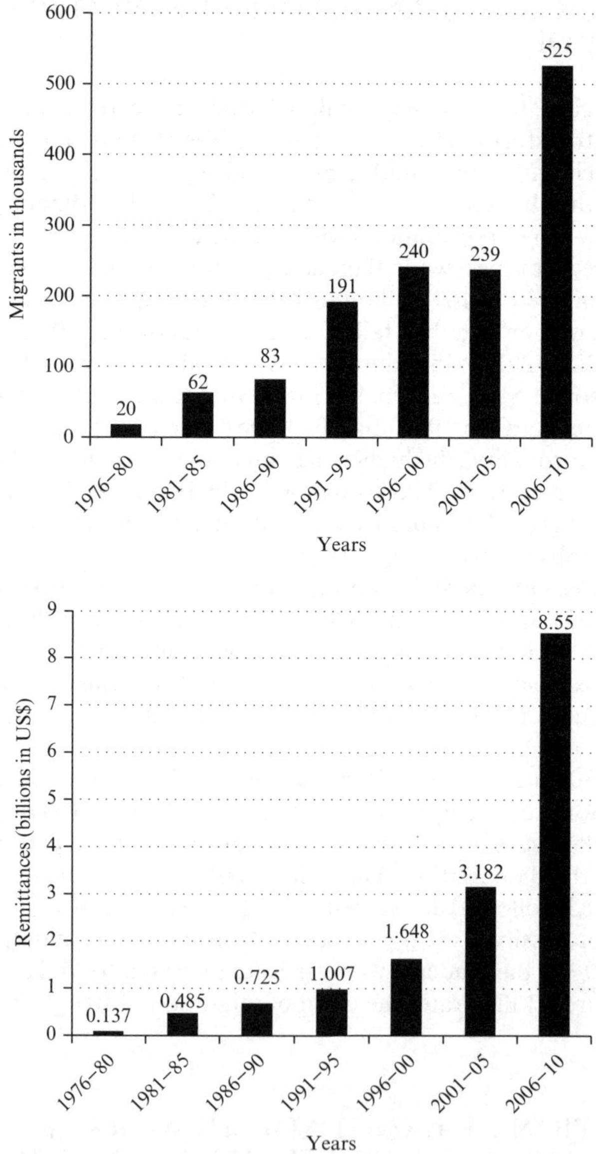

Source: Data compiled by the authors from BMET (2011).

Figure 9.1 Trends of migration of Bangladeshi workers (in thousands) and remittances (billions in USD) during 1976–2010 (five-year average)

decade. Since 2001, the number of Bangladeshi migrants to Malaysia increased to reach an average of 36456 per year during the period 2008–11 (BMET, 2011).

A variety of factors seem to influence choice of Malaysia as a preferred destination of Bangladeshi workers and these are: (i) availability of jobs commensurate with low level skills of migrants; (ii) religious and cultural affinity; (iii) physical proximity and easy access; (iii) aggressive marketing by middlemen; and (iv) lax border controls etc.

In terms of occupations, 86.6 per cent of Bangladeshi workers who migrate to Malaysia are unskilled, 13.2 per cent semi-skilled and 0.2 per cent skilled (Maxwell Stamp Ltd 2010). Semi-skilled and unskilled workers mainly work in the plantation and agricultural sectors, especially in jobs that many Malaysians regard as '3D' jobs (dirty, difficult and dangerous) (Robertson, 2008). A small proportion of Bangladeshi migrants also work in the construction, manufacturing and services sectors.

Categorization of Bangladeshi Migrants in Malaysia through Legal Perspective

In addition to classifying Bangladeshi migrants in Malaysia by vocation, they are also grouped in terms of their immigration/contractual status.

Bangladeshis adopt a variety of means and migration pathways. Some enter the country with proper contracts. Some enter with a visit visa and then seek a job. There are others that enter the country legally with a contract, become illegal at the expiry of the contract and continue staying in the country in the hope of securing a new one.

Therefore from the legal and contractual perspectives Bangladeshi migrants in Malaysia can be categorized as:

1. 'legal' migrants with jobs: those who enter the country with valid visas and job contracts and continue as 'legal';
2. legal migrants who become 'undocumented' or 'illegal': those who enter the country legally with jobs but due to weaknesses/complexities of rules and regulations and manipulation by employers become 'undocumented' aliens while still retaining jobs,[4] or who become illegal at the expiry of the initial contract but continue staying illegally with the hope or promise of getting jobs and the legalization of their stay;
3. legal migrants without jobs: those that enter the country with a valid visit visa but without job contracts, with the promise of jobs on arrival; and

4. illegal migrants without jobs: those that enter the country illegally with the hope or promise of jobs and the eventual legalization of their stay.

These classifications are important as they help in understanding better the challenges Bangladeshi migrants face in Malaysia.

Almost 50 per cent of the cumulative total of 700 397 Bangladeshi migrants to Malaysia from 1978 to 2011 (BMET, 2011), fall into the 'illegal' or 'undocumented' categories. There are also migrants who begin the journey to enter Malaysia illegally in the hope that once inside the country they will manage to get a job, but do not succeed in completing the journey itself. Some are apprehended in Thailand by the law enforcement agencies on their way and face imprisonment, some even die. So far very little research has been undertaken to record the fate of the 'illegal migrants without jobs'.

Institutions and the Processes of Migration: View from the 'Malaysia Village', Bangladesh

In order to understand various dynamics of the processes of migration – which include, among other things, aspects such as accessing information regarding migration, migration decision and the final journey – of Bangladeshi workers to Malaysia, this chapter has drawn upon the experiences of a migrating village in Bangladesh, referred to in this chapter as 'Malaysia Village', from where most migrants have gone or are aspiring to go to Malaysia.

'Malaysia Village' reveals in a practical way the 'push' and 'pull' factors of migration; the processes and governance arrangements that contribute to decision making regarding migration; the methods and modalities applied for recruitment of aspiring migrants; mobilization by the aspiring migrants of resources needed to pay for the cost of migration; the journey; the life of the migrants at the destination country; remittances and patterns of utilization of remittance money at the household and community level; and the economic and social impacts of migration, both positive and negative.

Malaysia Village

The experience of international migration of Malaysia Village,[5] located in the north-west of Bangladesh from where most migrants choose Malaysia as their destination country, is somewhat new.

This is a traditional Bangladeshi village where agriculture is the main source of livelihood. A total of 1727 people with 368 households comprising roughly five members per household, live in this village. The literacy

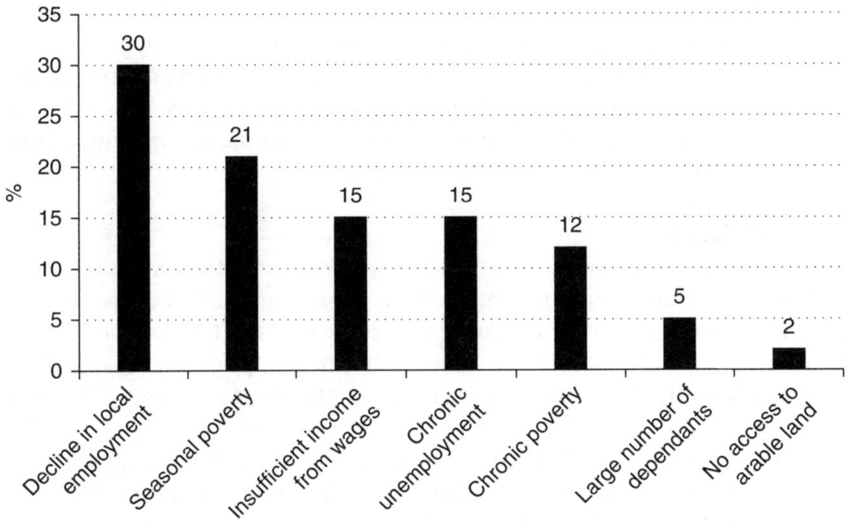

Source: Fieldwork, 2011.

Figure 9.2 Proportionate distribution of components of 'push factors' contributing to migration decision in Malaysia Village

rate of the village is around 31 per cent (male 37 per cent; female 25 per cent), which is significantly lower than the national average (60 per cent).

In recent times, the pressure of population and increased development activities, including infrastructure development (construction of roads, bridges etc.) have resulted in a steady conversion of large bodies of water, first into tracts of arable land and, subsequently, into homesteads and infrastructure development. These in turn have contributed to a significant decrease in the supply of land for agricultural activities, and this in turn has reduced agriculture-based employment opportunities in the village. These developments have also contributed to the rise of unemployment and poverty in Malaysia Village.

The 'push' factors of migration

Villagers in Malaysia Village report that an overall decline in livelihood options has contributed to the 'push factors', the motivation of migration in their village. Figure 9.2 reveals in detail the socio-economic factors such as 'declining local employment' (30 per cent); 'seasonal poverty' (21 per cent); 'insufficient wages' (15 per cent); 'chronic unemployment' (15 per cent); 'chronic poverty' (12 per cent); 'large number of dependents'

(5 per cent); and 'no access to arable land' (2 per cent), that have provided the backdrop or the key 'push' factors of migration in the village.

As stated earlier, the livelihood options have been contracted further by conversion of arable land to non-agricultural activities such as building of homesteads, construction of roads etc. that have also increased the risks of natural hazards such as flooding.

The 'pull' factors

The socio-economic compulsions, the so-called 'push factors' are not the only reasons that influence migration decision. The 'pull' factors, such as job opportunity, ease of travel and favourable immigration law of the host country including marketing by the recruitment agencies, also influence the decision regarding migration as well as the choice of the destination country.

As far as migrants of Malaysia Village are concerned, factors that seem to have motivated the villagers to migrate to Malaysia include (but are not necessarily limited to) the following: (i) 'influence of middlemen' (36 per cent); (ii) 'easy visa availability' (18 per cent); (iii) 'low cost of migration' (14 per cent); and (iv) 'more wages than home country' (13 per cent). See Figure 9.3.

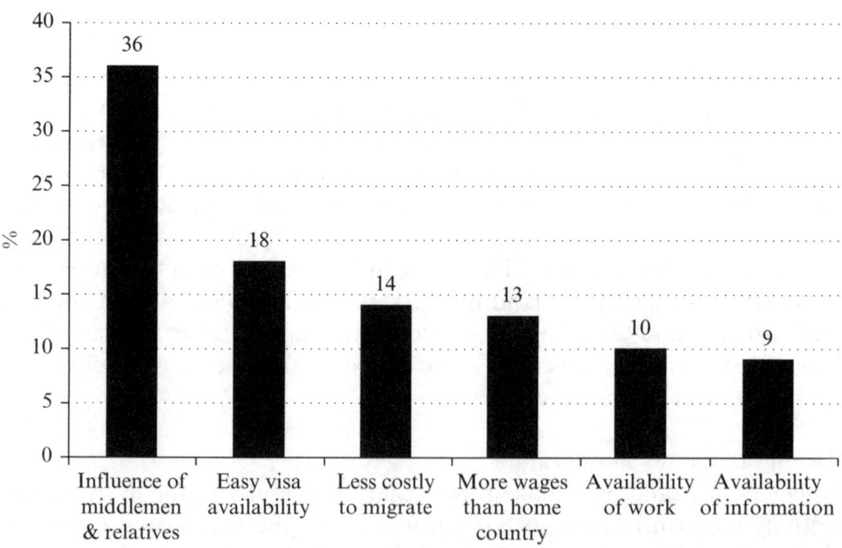

Source: Fieldwork, 2011.

Figure: 9.3 'Pull' factors that contribute to migration to Malaysia

It is noteworthy that the 'influence of middlemen' plays a far greater role (36 per cent) than that of the 'availability of work' (18 per cent) as a motivating or 'pull' factor of migration for the migrants of the study village. This is because villagers trust the networks of middlemen and past migrants, believing that these people, more than any other source, possess reliable information on and knowledge of job opportunities, wages, and working conditions in Malaysia, including information that is relevant to the processes involving migration. Furthermore, villagers also report that the 'agents' or the middlemen – some are former migrants themselves – assist in finding jobs and organize visas including all travel related arrangements for the prospective migrants, though over the years it has also been revealed that not all middlemen are trustworthy and that many villagers have fallen victim to their chicaneries.

BOX 9.1 EXPLOITATION OF AN ASPIRING MIGRANT BY MIDDLEMEN

Sobhan went to Malaysia for the first time in 1991, travelling through Thailand on a tourist visa, a low-box cost route. He was referred to a middleman by one of his relatives. Sobhan paid the middleman and was assured that even though the normal cost of travelling to Malaysia is more than USD 1250, he (the middleman) would organize the travel, part by land, part by air via Thailand, for USD 500. Sobhan bought the idea and travelled with others from his own village and from other districts to Thailand. They lived in a jungle without any provision for drinking water, sanitary latrine, food or proper accommodation. One night he lost consciousness and everyone thought that he had died. His family was informed of his 'death' by the middlemen but that ultimately proved to be false. Sobhan regained consciousness after he was given food and water. From Thailand he finally crossed a river by boat and reached Malaysia illegally. In Malaysia he was further exploited by both the middlemen and employers, economically as well as psychologically. None of the middlemen at any stage of the journey had delivered what they promised prior to migration (Sobhan, 45, grade 5, small businessman, returned migrant).

Table 9.1 Migration trends by socioeconomic classifications of migrants

Social class	Indicators	Number of households	Number of migrants
Extreme poor	Possess neither homestead nor arable land; work on others' land; begging	10	–
Poor	Have homestead land; possess no/small arable land; work on own and others' land; migration overseas occasionally	200	6
Lower middle	Have homestead land; own 0.3–0.6 acre of arable land; work on own and others' land; migration overseas	135	40
Upper middle	Have homestead land; 3 acres of arable land; work on own land; shallow machine operation; public sector worker; migration overseas	15	13
Rich	Have homestead land; 6 acres of arable land; shallow machine operation; public sector workers; migration overseas	8	2

Source: Fieldwork, 2011.

Processes Migration: Institutions, Networks, Agencies and the 'Middlemen'

A myriad organizations, formal and informal, and rules and regulations guide the processes of legal migration. However, not all migrations are legal and over time and due to complexities with rules and regulations, the status of some legal migrants shifts to become illegal. Understanding the processes of migration requires attention to both legal and illegal or informal processes.

Profile of Malaysia Village migrants

Constructed on the basis of villagers' self-assessments, Table 9.1 reveals that in Malaysia Village, of 368 households, 10 (3 per cent) are 'extreme poor'; 200 (54 per cent) are 'poor'; 135 (37 per cent) are 'lower middle'; 15 (4 per cent) are 'upper middle'; and 8 (2 per cent) are 'rich'.[6]

This table also maps the socio-economic status of the migrant house-

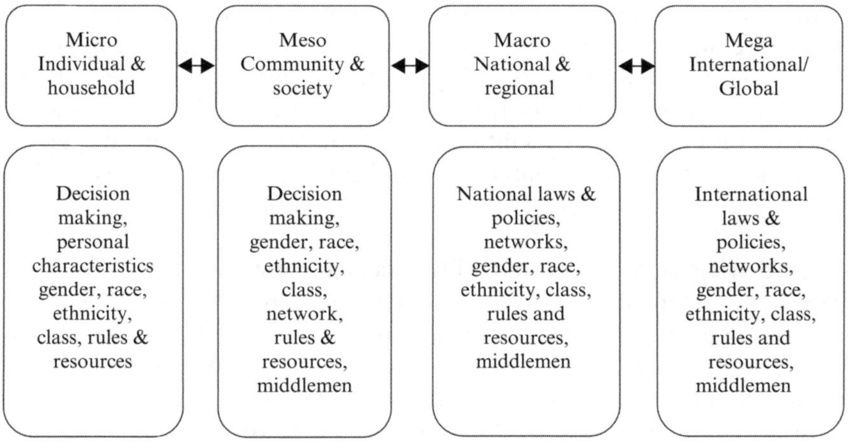

Figure 9.4 Agents and institutions of migration: four levels

holds, revealing that the 'lower middle' and 'upper middle' households who possess reasonable amounts of both financial and other assets (and thus have the capacity to invest in migration processes and face livelihood uncertainties) constitute the bulk (87 per cent) of those who have migrated/are migrating to Malaysia from this village.

The 'poor' households, who are worse off than the 'lower middle', 'upper middle' and 'rich' households, but are better off than the 'extreme poor', migrate in small numbers and they mainly do so by selling either their homestead land (the only asset they possess) and/or by borrowing money from neighbours and relatives and/or from moneylenders at a high interest rate.

Members of the 'extreme poor' households who are desperate for jobs are unable to migrate as they do not posses any asset that can be collateralized to obtain the necessary capital to fund the costs of migration.

The processes of migration

Figure 9.4 provides a framework that describes the range of agents and institutions that, one way or another, influence the migration process and have organized/organize migration of the aspiring migrants of Malaysia Village, to migrate to Malaysia, some legally and others not.

Figure 9.4 identifies four levels of institutions – mega, macro, meso and micro – that get involved in facilitating and/or influencing the processes of migration. Among these, while the mega and macro level institutions establish the policy aspects – the supply and demand nexus – of migration, the 'meso' and 'micro' level agents and institutions, both formal and informal, engage more directly in conveying information regarding migration opportunities, reach out to the potential migrants and assist in recruitment including the final journey to the destination country. In other words, in the migration cycle, it is the meso and the micro level entities that facilitate more prominently the entire cycle of migration, from initial contact to the final journey to the destination country.

Migration decision: the role of family, middlemen etc.

Migrants seldom make decisions to migrate completely on their own. A large number of agents – institutions, networks and agencies – both official and unofficial, at both the source of and destination for migration, operate in tandem to facilitate (or impede) migration.

Information obtained from Malaysia Village reveals that migration decisions are made at two levels: primarily at the household level, with some consultation at the community level, and sometimes independently, where the aspiring migrant himself is the head of household. However, even in this case, the migrant does consult family members.

Some 80 per cent of migrants (61) from Malaysia Village reported that they had been influenced or assisted by middlemen, commercial agents linked to the networks of international migration. Past and/or current migrants from the village, friends and relatives and agents of manpower exporting businesses based in Dhaka, the capital city and other regional cities, also operate as middlemen. Some middlemen are influential persons recognized politically at the village level.

Even though processing of migration through informal agents does not always yield positive outcomes – of those who were processed by the informal middlemen in Malaysia Village 40 per cent reported positive outcomes and 60 per cent not so positive, influence of informal middlemen in organizing migration remains a key feature of migration from the Malaysia Village.

Cost and sources of funding the cost of migration

With the increase in the number of Bangladeshis migrating overseas, the channels as well as the sources of funding the cost of migration have also expanded and diversified over the years.

As most government-approved migration agencies and financial institutions that are enlisted officially to facilitate migration are urban-based, with little or no outreach at the village level, most rural migrants, including those at the Malaysia Village, rely heavily on informal sources such as the middlemen, moneylenders etc. Furthermore, as the number of prospective migrants far exceeds actual demand, the middlemen and the moneylenders take full advantage of the situation, charging fees and interest rates that are two to three times higher than those approved by the government.[7]

It has been estimated that in Malaysia Village, each 'poor' and 'middle class' migrant household spends on an average USD2748 and USD2771 per head respectively for migration, though the total cost fixed by the government is USD1025. The differences between these two amounts are appropriated by the middlemen.

Due to total absence or inadequate sources of financing by formal sources, the migrants in Malaysia Village report that more than two-thirds (69 per cent) of those who fall into the 'poor' category of migrants arrange finance for migration from moneylenders and neighbours at a monthly interest of 10–15 per cent. A third (34 per cent) of middle-class migrants who possess assets engage their own resources in financing migration costs. Most middle-class migrants raise money from relatives and from personal savings and thus borrow rarely from informal commercial sources. In order to finance migration costs, the poor typically sell the little arable land they possess, the land that they would otherwise cultivate to support their families, thus making them heavily dependent for their future survival on the income from migration. In other words, the migrants who fall into the 'poor' category face destitution if they fail to secure the promised job. As will be described later, in such instances where the poor sold all of their assets to pay for migration and then ended up not getting the promised job, they have become destitute.

In sum, absent or inadequate provisioning of credit from official sources prompts most migrants to turn to the informal sources to fund migration costs at a level that is double the official rate. 'Middle class' categories of migrants who possess their own resources and rarely borrow from informal commercial sources are at an advantage. The poor are the most vulnerable and with no or little access to formal credit borrow at a high interest rate and pay much higher fees, and thus incur both higher initial costs and the cost of debt repayment at a high interest rate.

Governance issues

In order to assist migration and its processes, the governments of both Bangladesh and Malaysia have formulated, with limited success, numerous rules, regulations and institutions. However, a considerable body of literature reveals that due to lack of clarity and absence of transparency and accountability in the myriad rules and regulations that currently guide migration in both Bangladesh and Malaysia, Bangladeshi migrant workers face numerous challenges and suffer various forms of abuse and exploitation at the hands of the employers, the middlemen, the law enforcement agencies and others (Dameron, 2005; Devadason, 2011; Amarjit, 2007; Athukorala, 2006).

Yaw (2002), for example, argues that in Malaysia the migration policies that have been promulgated as an interim measure to tackle the unforeseen challenges of labour shortages in the plantation sector, have failed to administer a strategic vision that is efficient, transparent, accountable and fair. In this regard, the Asian Migrant Centre (2004) suggests that – due to complexities and glaring deficits associated with governance of migration and the collusive corrupt practices of some recruiting agents, immigration officials, police and employers – not only have welfare, security and safety of migrant workers been significantly compromised, but the inflows of illegal migrants and their exploitation have also increased.[8]

Experience of Malaysia Village reveals that discriminatory labour laws and lax monitoring of implementation of migrant contracts subject the workers, in this case the Bangladeshis, to lower wages than promised. Malaysia Village migrants have reported that they experience significant discrepancies between the original contract signed at the village and the new contract, known as the 'substitution contract' that is given to them on arrival. There is no government provision to monitor consistency between the original contract and the contract on arrival.

Migrant workers are also barred from joining local trade unions, leaving them with little or no bargaining option. In the event that migrants do join a union, they face termination of contracts and deportation. Furthermore, migrants report that even though the Passport Act of 1955 bars such practice, many employers seize passports and other important documents that support the migrants' immigration status upon arrival and as a result, the migrant workers get exposed to all sorts of abuse and harassment by the employers, the middlemen, and the law enforcement agencies. It has been reported that these predatory employers and some dishonest recruiters avoid prosecution by 'managing' the law enforcement agencies.

In order to alleviate these problems, the Government of Malaysia in recent times has introduced *i-Kad/Jalan-Kad* (a specialist identity card for migrant workers) to allow migrant workers to travel inside Malaysia as legal aliens without passports. However, to get this card, considerable time and money, expenses solely borne by the migrants and not the employer, are required and most migrants avoid getting the card as they cannot afford it. Some of the returned migrants in Malaysia Village reported that even the possession of this card does not automatically guarantee protection against harassment and exploitation by the immigration authority including the RELA[9] (*Ikatan Relawan Rakyat Malaysia* or Malaysian People's Volunteer Corps), a special force which oversees the migrant immigration status of aliens.

In summary, it is evident that the existing arrangements of governance of migration – the network of institutions, laws, rules and regulations that engage in and regulate migration processes, organize recruitment, manage migrant welfare and security and so on – are both inefficient and corrupt, and mostly function to disadvantage and not to assist the migrants. Furthermore, the *ad hoc* nature of various measures that have been introduced by the Government of Malaysia have either failed to tackle the challenges to their fullest measure or have lacked comprehensive implementation.

Despite these difficulties, Malaysia remains a favoured destination for the Malaysia Village migrants and this is due to a combination of factors – steady shrinkage of livelihood opportunities in the village, demonstration effects of successful cases and aggressive marketing by the middlemen.

OUTCOMES OF MIGRATION; THE DYNAMICS OF REMITTANCE, EXPENDITURE AND SOCIAL COST OF MIGRATION

In general, the outcomes of migration can be classified as:

1. Financial remittance: the money that the migrants send home at regular intervals and the impact it creates on the migrant household's socio-economic status and community development.
2. Social remittance: knowledge, values and attitudes that the migrants transfer from the destination country to the origin country/village and the social change induced, if any.
3. Non-remittance effects: unintended outcomes or social costs of migration.

Remittances by the Migrants and Pattern of Expenditure

On average, each migrant from Malaysia Village earns a total of USD196 per month in Malaysia (approximately 4.8 times higher than the average income for a worker in Bangladesh). However the net amount that finally reaches migrant households and is available for consumption/expenditure at this level depends on a number of factors, namely (i) the actual amount earned (this refers to any differences between the 'contracted amount' and the amount actually paid, and also between the amount actually received and the net amount after subsistence and other related costs including the transaction costs); (ii) the amount of on-going repayments of debts incurred to pay for migration related costs; and

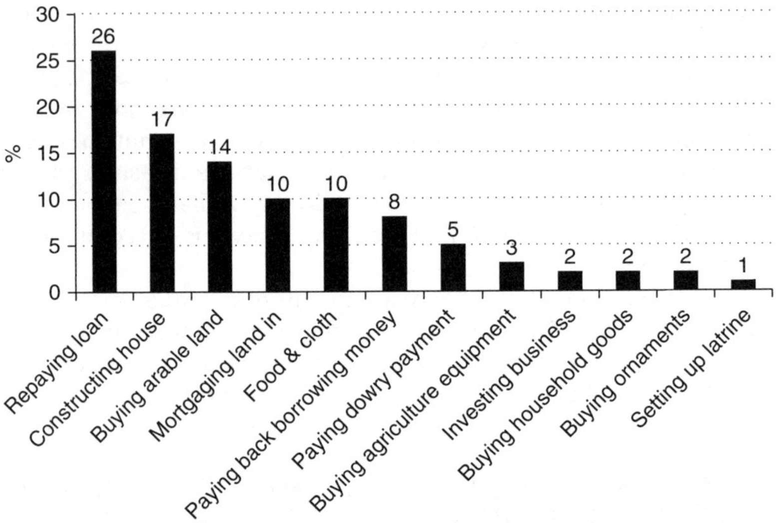

Source: Fieldwork, 2011.

Figure 9.5 Proportionate distribution of expenditure/consumption of remittances at household level in Malaysia Village

finally, (iii) the amount that is paid to middlemen and other rent seekers, at both ends. Moreover, remittances vary between first- and second-time migrants. Due to their ignorance of rules and regulations and of their rights and privileges, first-time migrants end up spending more on transaction costs and thus gain little from migration. The second-time migrants who have become more conversant with rules and opportunities are able to cut down on transaction costs and increase their net income significantly. Figure 9.5 presents an itemized proportionate distribution of expenditure of remittance money by the migrant households in Malaysia Village.[10]

It is evident from Figure 9.5 that on average, 'repaying loan' (26 per cent), 'constructing/repairing home' (17 per cent), 'buying arable land' (14 per cent), 'mortgaging land in'[11] (10 per cent) and 'food and cloth' constitute the five most important items of expenditure of remittance money at the household level. Observations in Malaysia Village also indicate that the poorer the migrant household, the higher the remittance expenditure on loan repayments followed by food, repairs to homesteads etc. For the 'poor' migrant household net impact on poverty, if at all, takes much longer to eventuate.

Table 9.2 Changes in socio-economic conditions of migrant households (before and after migration) in Malaysia Village

Social class before migration		Social class after migration			Reported improvement in socio-economic circumstances	
		Poor	Middle	Rich	Yes	No
Poor	21 (42%)	11 (52%)[a]	9 (43%)	1(5%)	14 (67%)	7 (33%)
Middle	29 (58%)	1 (3%)	25 (86%)	3 (10%)	12 (41%)	17 (59%)
Rich	–	–	–	–	–	–
Total	50 (100%)	12 (24%)	34 (68%)	4 (8%)	26 (52%)	24 (48%)

Note: [a] Percentages in brackets beside frequency counts are row percentages.

Source: Fieldwork, 2011.

Remittance and Change of Socio-economic Conditions of Migrant Households

In Malaysia Village, the impact of remittances on changes in the socio-economic conditions of migrant households varies from case to case. Grouping them by social class, Table 9.2 shows the impact on the status of socio-economic conditions of migrant households by pre-migration social class of migrants.

Table 9.2 indicates that 52 per cent of all migrants reported that they have experienced improvement in their socio-economic circumstances after migration. The pre-migration poor are more likely than the pre-migration middle class to report improvement in their socio-economic circumstances, and to report an upward shift in their social class position, after migration. However, the pre-migration poor are more likely, also, to report that their conditions had worsened due to migration.

Migrants who reported worsening socio-economic conditions are mainly first-time migrants who either have borrowed heavily to invest in migration-related transaction costs or were cheated of their legitimate wages. Their lower net income failed to service the debt fully and they were compelled to sell properties back home to repay debts.

Second-time migrants who become fully knowledgeable of the entire migration process and the rules and regulations do better and thus earn more remittance income and change household socio-economic status more convincingly. Motiur, for example, a returned migrant, reports that when he first migrated to Malaysia he was unable to contribute very much for his family. On his second attempt, however, he did well and improved, quite significantly, the family's socio-economic circumstances. Buchenau (2008) also confirms that second-round migrants increase their net income significantly.

As the net income earned and remitted by Malaysia Village migrants is somewhat modest, the contributions they make at the community level, for example, donations to religious institutions such as religious schools ('Madrassas') including mosques (79 per cent) and charity to the poor and other welfare activities (11 per cent) are also modest in amount.

In terms of social remittances, several returned migrants have reported that their migration to Malaysia taught them the values of hard work, punctuality and civic responsibility. At the same time, they also report that the technical know-how they applied or learned in Malaysia has been of little or no use back at home.

Non-remittance or Social Costs of Migration

There are several unintended outcomes or 'non-remittance' effects or 'social costs' of migration that many migrants experience. In Malaysia Village a significant social cost of migration is chronic debt. Although debt is a common issue for all migrants, poor and 'undocumented' migrants who borrow at a high interest rate from the moneylenders for migration spend most of their earnings on debt repayment and thus contribute little or nothing to their households. 'Undocumented' migrants incur significant transaction costs in Malaysia that, in turn, reduce the net income they earn and thus their remittances. Among the undocumented migrants, those from 'poor' households are especially likely to fail to make timely repayment of their debt. In order to settle that debt they often sell the only asset they possess – their homestead land – and, in the process, make their families homeless and destitute.

At a more collective level, overseas migration of large numbers of men from the village seems also to have contributed to seasonal labour shortages in Malaysia Village, forcing some migrant households to fill the gap by hiring labour from elsewhere, at a high cost. This has also contributed to the reduction of net income accrued by these migrant households.

Social costs

It is also evident that in Malaysia Village, the guarantee of a regular flow of remittance money has contributed to the rise of a number of social ills such as promotion of a 'culture of dependency' among the youths of the migrant households. Many villagers also report that easy money from remittances may have contributed to a range of socially unacceptable behaviours such as alcoholism, gambling, and extra-marital affairs at the village level. On many occasions these have led to conflict, tension, marital disharmony and divorce, and in some cases, breakdown of family authority and demise of extended family norms. The issue of control and management of migrant remittance money seems to have been a source of much conflict in several migrant households, contributing to, in some cases, breakdown of the traditional household power structures. In extended family households in Bangladesh, the father is regarded as the traditional head of the family and thus there is an expectation that he will receive and manage remittance money. However, in some migrant households, the non-observance of this norm (the spouse or another relative may have been entrusted with this task) has caused much tension and bitterness in the migrant household. Furthermore, some migrants who sell inherited land (an important symbol of social identity in a Bangladeshi village) to pay for migration costs, encounter loss of familial and other social ties and, in the process, lose social identity, a key element of a village existence in Bangladesh.

Other social challenges that arise from the prolonged absence of migrants (sometimes for periods of up to four or five years) include disempowerment and chronic depression of migrant spouses and also an increase in their workloads. For example, in the absence of their migrating husbands, many migrants' wives are forced to move back to the homes of their parents-in-law, contributing to the loss of whatever limited independence they enjoyed when their husbands were around. At the houses of their parents-in-law, these left-behind spouses are forced to take on an additional workload of the extended family household and in the process suffer not only physical stress but also emotional indignity.

Other social costs that accompany migration include a range of juvenile delinquencies including incidence of school dropout among the children of migrant households. In summary, the social costs of migration are many and these are rarely reported or assessed properly. The social costs of migration such as chronic debt and destitution (especially for the poorer migrants), family breakdown, disempowerment of the 'left behind' migrant wives, an increase in hedonistic behaviour and in school dropout rates, beg the question whether migration indeed is a net contributor to the overall well-being of migrant households.

SUMMARY AND CONCLUSIONS

The study stresses that notwithstanding its many challenges, the twin phenomena of rapid economic growth accompanied by a decelerating birth rate in some Asian countries and concurrently, relatively slower economic growth, rising social and economic inequities and increasing population in others, combined with easy access to and cheaper transportation costs of travel, combined with cultural affinity, has made Asia to Asia migration – the flow of migrant workers from less developed to more developed Asian countries – an important feature of Asia's future.

The Malaysia Village experience has demonstrated that the majority of migrant households have benefited from migration, especially financially. However, accounts from Malaysia Village have revealed that the amount of net financial benefit earned through migration depends, on the one hand, on the transparency and integrity of the migration processes and, on the other, on the socio-economic background of the migrant. The richer the migrant, the better the income from migration remittance, and the opposite is the case for the poor. Furthermore, the Malaysia Village experience is that lack of government outreach accompanied by exploitation of migrants by the middlemen that dominate current migration processes and their governance at home and also in Malaysia, significantly reduce financial benefits, jeopardize migrant safety and aggravate the social and psychological consequences of migration.[12] Foregoing discussions have also recorded several social costs such as breakdown of family norms, gender disempowerment, increases in school dropouts and deviant social behaviours that accompany migration.

Also in A2AM, a less studied issue is that of the relationship between migrant welfare in the destination country and migrant productivity. As this chapter has not examined this particular aspect in any depth it is unable to state what the productivity losses are, if any, to the Malaysian economy caused by weak protection of migrant rights and privileges, especially those of the Bangladeshis. This could be the subject of future research.

In sum, Asia to Asia migration is a fact of life and is likely to continue for some time. However, in order to make this emerging regional phenomenon a win-win situation, special attention must be given to a range of issues that seem to reduce the benefits of migration. These include among other things: better and more accountable management of migration processes (from information sharing to recruitment); access to affordable funding for bearing the costs of migration especially for the poor (for example, provision of microcredit for the poorest of the poor aspiring migrants); and provision of counselling to alleviate the social costs of migration.

Furthermore, better protection of migrant rights and their welfare is not merely an issue of human rights (which is nevertheless important) but is key to bolstering their productivity and thus the economy of the destination country. The receiving countries of Asia, including Malaysia, must address these deficits more efficiently and accountably.

Finally, given that benefits of migration, especially at the household and community level, are at best mixed, and also that migration entails significant social cost, should not a sending country such as Bangladesh think long term of development strategies that are self-sustaining, equitable and employment generating? After all, existing evidence suggests that countries that rely on migration for development and poverty alleviation have achieved neither.

NOTES

1. A2AM excludes the Gulf states of the Middle East, which since the oil boom of the mid seventies have experienced significant in-migration of Asian labour.
2. All figures rounded.
3. Unskilled workers include housemaids, agricultural workers, hotel workers, and such basic labourers as cleaners, cart loaders, and cotton pickers; semi-skilled workers include tailors, gardeners and masons; skilled workers include manufacturing or garment workers, drivers, computer operators, electricians, mechanics, welders, painters, cooks, plumbers; and professional workers include doctors, engineers, teachers, nurses, accountants, computer programmers, pharmacists, foremen, diploma engineers, paramedics, sales personnel etc. (BMET, 2011).
4. This relates to those that are made to surrender passports on arrival to their employers and thus live in the country as what many term 'undocumented' aliens.
5. The expert study reported here is drawn from a broader, comparative study of two migrating villages in Bangladesh. Two periods of fieldwork were undertaken in 2011 and 2012. Data were gathered through household surveys, observation, focus group discussions and in-depth interviews with migrants, migrants' family members, and local officials in the two villages. Malaysia Village is a pseudonym used to protect the identity of research participants. The other migrating village, 'UK Village' lies in the north-east of Bangladesh. UK Village, by contrast to Malaysia Village, has a near one-hundred year long history of migration, mostly to the UK. The comparative dimension of the overall study has been important in highlighting and demonstrating how differing socio-economic and governance dynamics of migration produce different outcomes.
6. In a focus group session of older male persons in the village, participants were asked how they would classify households in terms of socio-economic indicators. The internationally reputable microcredit institution, Grameen Bank, defines 'poor' as those who own less than 1.0 acre of arable land.
7. Loans to fund migration are sometimes backed by a mortgage on agricultural land. The borrower does not have to pay interest in these cases as control of the land is transferred to the lender who benefits from its output until the loan is repaid. This arrangement seems to be common in most rural areas.
8. These practices became more pronounced when the Malaysian government launched the system of the 'Outsourcing License' (OL) in 2006. Under this system, the companies that recruit less than 50 foreign workers were required to apply for foreign labour to the outsourcing companies that were tasked to manage recruitment on their behalf. Although the OL was introduced to eliminate middlemen or agents that till then exploited the workers during recruitment, the delay and sometimes high cost involved in obtaining approval from the higher authority discouraged the companies, which

are always in a rush to recruit. This resulted in their employing 'illegal' workers who migrated through the middlemen (POEA, 2006).

9. RELA Corps is a paramilitary civil volunteer corps formed by the Malaysian Government. Their main duty is to check the travel documents and immigration permits of foreigners in Malaysian cities, including tourists, visitors and migrants to reduce the increasing number of illegal immigrants in Malaysia. RELA has the authority to deal with situations like policemen, such as raiding suspected streets or places such as factories, restaurants and even hotels. They are also fully authorized to conduct the interrogation and even detaining of people who do not have their travel documents, such as passports and/or working permits.
10. A recent study has revealed that migrants spend one third of income for family at home, one third for debt repayment and one third for self on subsistence and other incidentals (Buchenau, 2008).
11. Under 'mortgaging in' arrangements, migrants' families use remittances to obtain use of arable land in return for interest-free loans to landholders for a fixed period. When loan monies are repaid, use of land is returned to the landholders.
12. Field research from 'UK Village' – a migrating village in the north-east of Bangladesh where most migrants go to the UK through networks of family and friends – has shown that its migrants experience low levels of abuse and exploitation and thus the returns in terms of financial remittances are much better. Labour rights of UK village migrants also are better protected in the UK where, compared with Malaysia, governance arrangements are better and conducive to better human rights. Another study (Buchenau, 2008) has also confirmed the positive relationship between the use of family/friends networks in migration and migration outcomes.

REFERENCES

Amarjit, Kaur (2007), 'Migration matters in the Asia-Pacific region: immigration frameworks, knowledge and national policies', *International Journal on Multicultural Societies*, **9** (2), 135–57.

Asian Migrant Centre (2004), *Asian Migrant Yearbook 2004*, Hong Kong: Asian Migrant Centre.

Athukorala, P. (2006), 'International labour migration in East Asia: trends, patterns and policy issues', *Asian Pacific Economic Literature*, **20** (1), 18–39.

Buchenau, Juan (2008), 'Migration, remittances and poverty alleviation in Bangladesh', United Nations Development Program, Preparatory Assistance for Promoting Pro-Poor Trade (PA-PPT), Dhaka, Bangladesh.

BMET (Bureau of Manpower, Employment and Training) (2011), 'Overseas employment and remittances from 1976–2011', BMET, http://www.bmet.org.bd/BMET/viewStatReport.action?reportnumber=30.

CIA (Central Intelligence Agency) (2011), *CIA World Factbook 2011*, http://www.indexmundi.com/g/g.aspx?c=bg&v=74.

Dameron, Kirk (2007), Review of Robert E.B. Lucas, *International Migration and Economic Development: Lessons from Low-Income Countries*, in *Journal of Economic Issues*, **41** (1), March, 291.

Devadason, Evelyn (2011), 'Policy chaos over migrant workers in Malaysia', East Asia Forum, Economics, Politics and Public Policy in East Asia and the Pacific.

Hugo, Graeme (2005), 'Migration in the Asia-Pacific region', paper prepared for the Policy Analysis and Research Programme of the Global Commission on International Migration, National Centre for Social Applications of GIS, University of Adelaide.

Maxwell Stamp Limited (2010), *Study on the International Demand for Semi-skilled and Skilled Bangladeshi Workers*, TVET Reform Project, Bangladesh.

POEA (Philippines Overseas Employment Administration) (2006), *POEA Market Update: Malaysia Adopts Outsourcing System in the Hiring of Foreign Workers*, series no. 13, Marketing Branch, http://www.poea.gov.ph/lmi_kiosk/market_update.htm, retrieved 10 January 2012.

Robertson, Philip S. (2008), 'Migrant workers in Malaysia – issues, concerns and points for action', Fair Labour Association, Washington, USA.

United Nations (2004), *World Economic and Social Survey 2004: International Migration*, Department of Economic and Social Affairs, New York: United Nations.

World Bank (2007), 'Remittance trends 2006,' Migration and Development Brief 2, http://www.worldbank.org/prospects.

World Bank (2011), *Migration and Remittances Fact Book* 2011, 2nd edn, Washington DC: World Bank.

Yaw, A.D. (2002), 'Introduction: migrant workers in Pacific Asia', in A.D. Yaw (ed.), *Migrant Workers in Pacific Asia*, London: Frank Cass, pp. 1–18.

PART III

Realizing the Asian Century: The Challenges

10. Sustainable resource development in Asia: challenges and opportunities

Vigya Sharma and Tapan Sarker

INTRODUCTION

This chapter examines the role that mineral resources and sustainable mining may play in Asia's future growth and development. Across Asia, several big and small nations are endowed with rich deposits of valuable metals and minerals. While for some of these nations, mining has remained a long-term economic activity, for other smaller nations, minerals development is slowly but surely catching momentum. This chapter has a twofold purpose: firstly, it reflects on the experiences of other resource-rich regions in the world and in doing so, suggests how resource development can provide long-term opportunities to enhance the socio-economic status of Asia's developing regions. Secondly, it acknowledges the transformational nature of mining from a socio-cultural and ecological standpoint. A key objective of this chapter, therefore, is to understand how Asia's current and future policymakers might best optimize opportunities provided by mining without succumbing to the many potential challenges – social, environmental, institutional as well as economic – inherent to resource development.

The chapter is organized in six parts. The following section 'Sustainable Resource Development: Principles and Practice' provides a theoretical underpinning of the key principles and characteristics identified with sustainable resource development. The section 'Resource Development in Asia: An Intellectual Review' reviews the state of mining in Asia and identifies the region's vital resource sector players. In the section 'Responsible Mining in Asia: Challenges and Opportunities for Development', we discuss current and future pathways for Asia's resource development process and highlight potential opportunities and challenges that may enable or impede long-term holistic development across Asia. The section 'Mining for Development: A Process of Learning, Sharing and Knowledge

Building', draws together experiences of select resource-rich regions in the world and analyses how lessons learnt from these experiences may help influence and shape the future of resource development in the Asian century. The final section summarizes the chapter's key findings.

SUSTAINABLE RESOURCE DEVELOPMENT: PRINCIPLES AND PRACTICE

This section introduces the concepts of 'sustainable resource development' and 'responsible mining' as interpreted by international institutions such as the World Bank and the International Council of Mining and Metals (ICMM). In doing so, it highlights two key ideas in this space – 'social license to operate' (SLO) and 'corporate social responsibility' (CSR) – to provide a useful theoretical lens to analyse Asia's resource development trajectory as discussed in subsequent sections of this chapter.

As a transdisciplinary concept, sustainability is prone to varied interpretations based on scale, multiplicity of domains and socio-cultural and institutional contexts (Martens 2006). The conceptualization of sustainability from a resource development perspective is quite similar in that it is vulnerable to a range of external and internal factors including, but not limited to, global economic performance; local context, defined largely by community needs and institutional arrangements; past resource extraction and industrial legacies; sensitivity and resilience of ecosystems to change; and the interaction of mining with existing social, economic and environmental characteristics defining the region. Esteves (2008, pp. 40–41) rightly captures the essence of the ongoing debate on sustainability in a resource development setting:

> Sustainability has different implications 'depending on which side of the bulldozer you are on'. As such, it is a societal construct where multiple paradigms exist . . . Mining presents dichotomies, at one extreme, opportunities for the production of substantial wealth, and on the other, threatening destruction of those lands and associated community life.

There is no doubt that a number of characteristics inherent to extractive industries result in both barriers and opportunities for operationalizing sustainability at the local and regional level (Gibson and Klinck 2005). Over-dependence on natural resources for social and economic development, often leading to dysfunctional governments and communities is a fundamental issue of concern in a typical mining context. As a consequence, numerous other challenges emerge, which can influence sustainability pathways in mining regions in important, at times irreversible, ways (Ross 1999; Sachs and Warner 2001; Robinson et al. 2006). The list below – while not being exhaustive – identifies several such challenges that threaten holistic development of resource-dependent institutions and communities:

1. Long-term changes to labour and employment patterns in the region, thereby limiting opportunities for sustained development of other forms of non-mining economic activities, such as agriculture (Richards 2009);
2. Increased inflow of transient workforce with varied commitment to the region, leading to loss in community solidarity and belongingness to place (Lawrie et al. 2011);
3. Income inequality, resulting in further disadvantage for those engaged with non-mining activities (Rolfe et al. 2007);
4. Social and ecological inequity more generally, whereby unequal regimes not only lead to inconsiderate exploitation of resources, but the patterns of redistribution of returns are also skewed over the longer term (McNeill et al. 2012);
5. Loss of attachment to ancestral and cultural land, and associated impacts on indigenous culture and communities, thereby impacting psychological wellbeing (Ritter 2001; Katona 2002; Ballard and Banks 2003);
6. Strong gender bias in mining due largely to remote location of the industry combined with weak gender-inclusive nature of the work environment (Solomon et al. 2008; Hill and Newell 2009; Kemp 2009);
7. A range of short- and long-term ecological impacts, including pollution of water sources, increasing levels of dust and noise, deforestation and loss of biodiversity, land degradation, and challenges with tailings storage (McMohan and Remy 2001; Aubynn 2003; Kitula 2006; Franks et al. 2011); and
8. Finally, often neglected or poorly managed post-mining impacts, while extremely critical for the future sustainability of the region, result in mining legacies with lasting consequences for local communities, ecosystems and institutions (Solomon et al. 2008).

In light of the above challenges, what then might sustainable resource development entail? The ICMM principles for sustainable development in mining provide useful pointers as they reiterate the importance of fundamental building blocks of sustainable development advocated by the Brundtland Report (WCED 1987), the pioneers of the concept. According to the ICMM (2011), the presence of ethical business practices that manage risks while upholding health and safety of those involved, the presence of a sound governance apparatus that values varied cultures and respects human rights, a consistent focus on maintaining robust ecological health, thus conserving biodiversity, and a system of transparent and accountable stakeholder engagement are necessary pre-conditions for resource development to be sustainable from a holistic standpoint.

Sustainable resource development thus necessitates a systematic approach to deconstruct the process of impact creation and dissemination, to help generate positive value for mining regions and communities on an ongoing basis (Brereton et al. 2008). In recent times, the dynamics of interaction between various actors and disciplines in a resource development setting has witnessed gradual transformation: from being purely need-based and 'a necessary condition of the acceptance . . . whether by the host government, a regional authority, the community at large or local interests' (Brereton and Forbes 2004; Crowson 2009a, pp. 106–8; Kemp 2009), mining has embraced a more encompassing attitude towards appreciating the 'shared value' generated in the process for all stakeholders – local and regional governments, industry and communities.

In addition to the current knowledge established around the concept of sustainable resource development, a relatively recent idea of 'responsible mining' adds a further layer of institutional robustness to ensuring that mining contributes to positive development outcomes. Responsible mining advocates that 'the rights of all stakeholders, especially of local people' are acknowledged and respected (Browne et al. 2011). It applies a lens of eight basic principles, brought together from varied discourses such as political economy, ethics and justice, sustainable development, governance and industrial ecology. These principles are presented in Figure 10.1. The concept of 'responsible mining' is particularly recent to resource-rich developing regions of Asia. It has been observed that a methodical application of these principles positively influences mining's contribution towards holistic development and poverty reduction, while also ensuring that economic prosperity does not threaten deterioration in the ecological and socio-cultural quality of life (Browne et al. 2011).

With 'responsible mining' strengthening the potential of positive outcomes from resource development, concepts such as CSR and SLO provide further means to that outcome. They serve as pragmatic tools to help the mining industry embrace the idea of sustainability into strategic decision-making through the project lifecycle (DRET 2011). On the one hand, the terms and conditions around CSR are often dictated by statutory regulations to formulate a process that ensures mining companies 'take account of [their] economic, social and environmental impacts in the way [they] operate – maximising the benefits and minimising the downsides' (Owen 2007, p. 1). On the other hand, SLO is a rather abstract phenomenon that

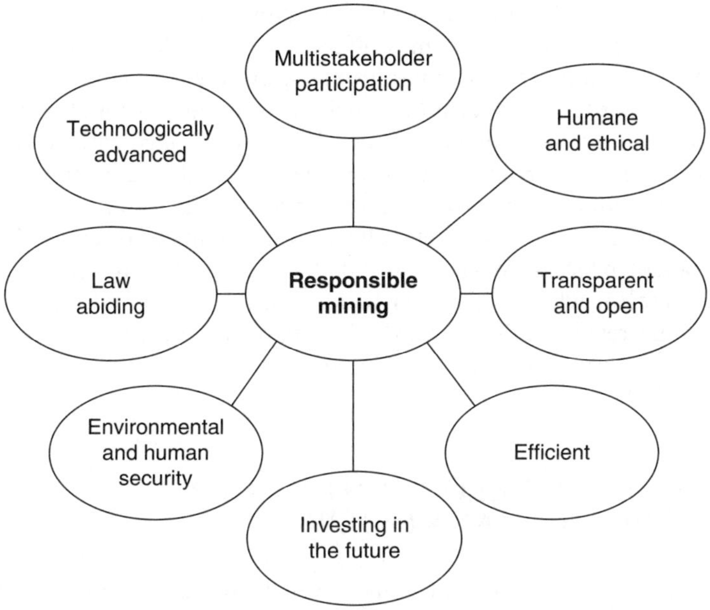

Source: Browne et al. (2011).

Figure 10.1 Principles of responsible mining

maintains a strong voluntary flavour to its conceptualization as well as execution. In its most basic form, SLO entails an evolutionary process by which local communities and other stakeholders can discuss their values and beliefs to help mining operations not only work towards further strengthening these values but also prevent them from undertaking actions that may undermine local stakeholder priorities and concerns. Although the state of a 'social license' *per se* may change over time and amongst different stakeholder groups (Black 2010), it remains a pragmatic tool that encourages the mining company to 'make informed decisions in an open and transparent process that includes comprehensive public participation and input' (Thomas and Nowak 2006; Shepard 2008, p. 2). Doing so, in turn, helps formalize a process of voluntarily 'sharing responsibility' with other actors including the government and local communities in building dynamic, progressive and sustainable societies.

Finally, this chapter frames sustainable resource development in the concept of 'shared responsibility' while acknowledging that varied priorities, ambitions, and objectives may continue to persist simultaneously across different stakeholder groups (Guerin 2009; McPhail 2009).

Applying the construct of sharing responsibility for both individual and cumulative actions and outcomes offers two particular advantages: first, it allows understanding sustainability as a more encompassing idea that considers long-term dimensions, such as intergenerational aspects of equity in resource management. Second, it helps engage with sustainability as an iterative, ongoing process that is influenced by the complex – and dependent – interplay between social, environmental, economic and institutional interests of different stakeholders in the minerals extraction and development process (Waye et al. 2009). For developing regions within Asia, several of which are fast-growing economies, joint stewardship of, and owning responsibility towards the use of natural resources can thus play a particularly important role in effectively transforming natural into social capital.

RESOURCE DEVELOPMENT IN ASIA: AN INTELLECTUAL REVIEW

In this section, we analyse the trends as well as the current state of resource development in Asia with a particular focus on Southeast Asia. With increasing globalization and growing demand for natural resources both within and outside Asia, resources development projects have become an important source of investment in Asia in recent years. Such projects contribute significantly to the economic development of many Asian regions by generating significant government revenues, foreign exchange earnings, employment opportunities, enhanced technological innovation, and opportunities for foreign direct investment (FDI) and private sector development. However, as discussed in the previous section, these economic enhancements across Asia have been accompanied with serious environmental, health and social challenges, often in already disadvantaged and remote parts of the continent.

In discussing Asia's past, current and future scenarios with respect to resource development, three important questions emerge that call for further discussion. First, what are the primary reasons for an upward trend in resource development in Asia? Second, are the factors that propelled rapid resource development in Asia likely to continue in the future? Finally, and more importantly, driven by the prospect of higher revenues, resource-rich developing countries in Asia are continuing to exploit their natural resource endowments to fuel further growth and new economic development opportunities. A key challenge then remains: how successful have resource development exercises in Asia been thus far in achieving long-term holistic development?

As this chapter will demonstrate, there are no direct or simplistic answers to these questions. On the one hand, it is clear that rising demands for mineral resources from fast growing markets in Asia (particularly China and India) and a surge in mineral prices in recent years have contributed to past and ongoing efforts in some of Asia's remotest and least developed regions to maximize growth and development opportunities from resource development. On the other hand, a combination of important structural, institutional and development challenges including, *inter alia*, weak and ineffective institutions; disempowered, often mis-informed communities; negative legacies of resource development across different locations; over-exploited and degraded ecosystems; and social and institutional inexperience with resource development in some of Asia's newly emerging mining heartlands currently exist that call for urgent attention (Sachs and Warner 2001; Cronin and Pandya 2009; Carbonniery et al. 2011; Havro and Santiso 2011; Spiegel 2012; Mining Communications February 2008).

Resource Development Trends in Southeast Asia: Established and Emerging Markets

Southeast Asia is a region rich in forest, mineral, and water resources. Yet, the abundance of resources remains under constant scrutiny in light of severe pressure from population growth, ongoing destruction of watersheds, river and land pollution resulting from rapid urbanization, mining, and hydropower development, high degree of dependence on subsistence agriculture and low levels of socio-economic development. Among the countries in Southeast Asia, Indonesia presents an excellent case where the presence of natural resources, especially minerals, oil and forestry, played a major role in its economic development. Among its huge mineral endowments, tin, nickel, copper, gold, and coal, produced from world class deposits, are important commodities for both export and domestic markets. A total of US$12 billion worth mining investment was recorded in Indonesia until 2000. The development of mineral resources has since made a significant contribution to the economy, accounting for approximately 3.5–4 per cent of GDP between 2001 (OECD 2002) and 2009 (PricewaterhouseCoopers 2011). The relatively slow growth in Indonesia's natural resources sector between 1999 and 2009 was a consequence of the country's own institutional challenges, most notably generated by the wave of decentralization during this period (refer to sub-section 'Past experiences of resource development with weak institutions, poor governance and human rights abuse: implications for Asia's future' for more detail). In recent years, the mining industry's contribution to the overall

Indonesian economy has increased significantly (approximately 6.2 per cent in 2011), largely owing to high commodity prices and increasing demand for export (Indoterra 2011; PricewaterhouseCoopers 2011; Purba 2008).

For other countries in Southeast Asia such as Vietnam, Cambodia, and Lao PDR, resource extraction and development has emerged as an area of institutional priority in recent years (Lyday 2002; Oum 2009; Sothea 2010; DHVP Research 2011; Mining Communications 2008). Mining has undergone rapid growth over the last two decades in several parts of Southeast Asia. For instance, in Lao PDR, natural resource contribution to GDP has risen to 10 per cent in 2010, as compared to only 0.5 per cent in 2001. Similarly, mining production rose by 19 per cent in 2010. Table 10.1 provides a detailed analysis of the significance of extractive industries to the socio-economic development of several Southeast Asian regions.

Asia's Primary Mineral Consumers

There is little doubt that China's rapid growth and economic expansion fuel the continued extraction and development of Southeast Asia's natural resources. Since major policy reforms in the 1990s, both China, and to a lesser extent India, have become key consumers of minerals produced from the Southeast Asian region (Cronin and Pandya 2009). Despite their own significant production of coal and minerals, China and India together constitute the largest market for coal, copper, and other minerals, as well as rubber and edible oils from Southeast Asia. Figures 10.2 and 10.3 provide the total value of natural resource-based commodity imports by China and India, along with source regions, primarily from within Southeast Asia.

Finally, how resource-rich regions within Asia respond to development challenges will not only influence their own socio-economic development over the coming decades but will also have several important implications for those dependent on their mineral wealth both within and outside Asia. As much as large mineral endowments are often considered important preconditions to stimulate growth in several of Asia's emerging economies, the ultimate tenacity to translate this mineral wealth into sustainable resource development on a long-term basis will be tested by the presence of numerous challenges of social, environmental and institutional nature. The next section discusses further some of the challenges likely to affect Asia's resource development trajectory over the coming years.

Table 10.1 Socio-economic significance of resource development in Southeast Asia

		Indonesia	Laos	Cambodia	Vietnam	Philippines
Mining contribution to GDP (%)	2001	15	0.6	0.2	6.6	Early 1990s: 30% to ~2% in 2002
	2008–10	10 (regional contribution, much greater)	10 Annual growth rate > 10%	0.0057, but set to rise drastically	~10–11	~3
			Mining revenues as proportion of total govt revenues (<2% in 2005/06 to ~14% in 2008/09)	104 licenses issued since 2006 Investment up from $4 million (2008) to $11 million (2009)		
Production value US$		$125 billion (by 2014)	From $8 million (2002) to $800 million (2007)	–	$2.98 billion (current prices 2000)	From $2.6 billion (2010) to ~$3.7 billion (by 2015)
Total mining employment	2006–08	960000	30000 (~5% of non-agriculture employment)	21000	~300000	100000
Export significance of the mining sector		21 billion (72% of total coal exported)	40% of total exports	–	8.5 billion (2009)	2.5 billion (2007)

Sources: Fast Market Research (10 July 2010); Wu (2001); Lyday (2002); The Wallace (2003); Wu (2003); Oum (2009); Sothea (2010); DHVP Research (2011); PricewaterhouseCoopers (2011); GSO (2012); Purba (2008); The Jakarta Post (2012); Browne (n.d).

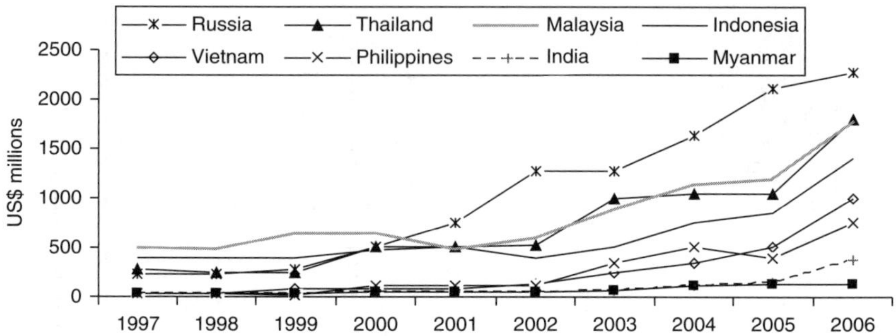

Sources: UN Comtrade Database (www.comtrade.un.org/db); Cronin and Pandya (2009).

Figure 10.2 Value of mineral resource imports to China (by exporting country)

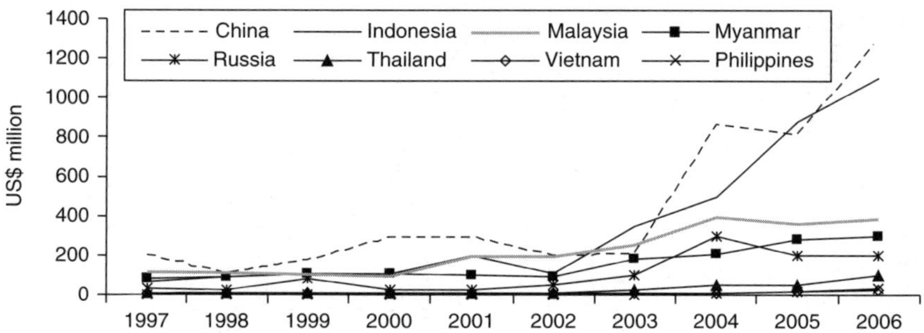

Sources: UN Comtrade Database (www.comtrade.un.org/db); Cronin and Pandya (2009).

Figure 10.3 Value of mineral resource imports to India (by exporting country)

RESPONSIBLE MINING IN ASIA: CHALLENGES AND OPPORTUNITIES FOR DEVELOPMENT

In light of the discussion so far, it is now established that several developing countries within Asia have either hosted mining activities for a long time (such as Indonesia and the Philippines) or are currently at the cusp of a large mining boom (for example, Laos, Cambodia and Mongolia). Owing to widespread diversity in social, economic, institutional and ecological landscapes across these regions, it is particularly complex to examine the nature and scope of factors across different scales – local, sub-regional and national – that may both challenge and influence the future course of resource extraction in Asia. Having said that, there are two particular issues that may threaten the future of resources industry over the long term (Slack 2009): (a) ongoing trend of extreme natural climatic variability combined with irrevocable human-induced climatic changes; and (b) distorted political and institutional agendas and structures that raise doubts over current institutional ability to uphold the principles of responsible mining in Asia. The chapter discusses these issues by drawing together experiences from three case studies across Asia.

The rationale for choosing these cases is twofold: firstly, while all three cases (Mongolia, Philippines, Indonesia) discussed here bring with them a suite of context-specific social, environmental and economic conditions that influence and shape their future sustainability pathways, from a purely institutional perspective, each case is characterized by a unique combination of socio-ecological and governance challenges that may potentially constrain the degree to which mineral resources are able to contribute to regional sustainable development. Secondly, these cases provide a systematic representation of opportunities and challenges arising across the various stages of the minerals development lifecycle. While Mongolia is a recent player in minerals extraction and development, Indonesia, on the other hand, has a far more established minerals sector and a number of negative and positive experiences in mining that can inform not only its own future socio-political course in mineral development but also provide valuable lessons to other emerging mineral-rich economies within and outside Asia. The mining industry in the Philippines – possibly the only region in Asia where a strong mining boom for many years was followed by a steep decline in both quality and quantity of mining projects – provides yet another perspective to better understanding resource development challenges and opportunities (Stark et al. 2006).

Resource Development in a Changing Climate: The Case of Mongolia

Changes to the natural climate may not only disrupt mining operations (for example, via losing access to mine sites, ports and key infrastructure such as roads and machinery) but also affect a large proportion of natural resource-dependent populations in Asia's many developing regions. Mining is an extremely water-intensive industry, and both the quantity and quality of water remain paramount to its long-term viability as an economic activity. With changes to climatic conditions projected to lead to increasing frequency and intensity of extreme weather events such as floods, cyclones and droughts, water availability and quality in remote mining regions remains a contentious issue (Slack 2009). Furthermore, some of the world's most mining-intensive regions are characterized by remoteness, extensive socio-economic dependence on natural resources for daily subsistence (for example, agriculture, fishing and pastoralism), and generally low levels of human development (Slack 2009). The interaction between ongoing livelihood struggles and increasing conflict over sharing common resources, particularly water, may generate resentment against the industry, thereby threatening its SLO or lead to more organized violent conflicts or both.

Mongolia provides a particularly timely example of the interaction between harsh climate, increasing institutional focus on resource development and a deepening crisis concerning the long-term viability of its traditional livelihood sources. From being a country where 'known reserves [we]re modest and . . . [did] not enjoy a superabundance of currently exploitable natural resources' a decade ago (Fish 2001, p. 325), Mongolia's strategic and market value as a result of the vast resource of valuable metals and minerals sitting underneath its expansive landmass has grown enormously at the international scale. It is now widely recognized as a country with vast resources of copper, coal, gold, fluorspar, uranium and other natural resources, which are expected to increase foreign investment in the country by up to US$25 billion by 2016 (FlorCruz 2011). Mongolia's GDP grew by more than 7 per cent in 2010, a direct result of large-scale mining investments, particularly in the ecologically sensitive South Gobi region (FlorCruz 2011). The minerals sector already employs in excess of 14000 people and contributes to more than 80 per cent of the national export earnings. Oyu Tolgoi in Mongolia's South Gobi region is the world's largest copper mine. Once operational in 2013, it alone is expected to contribute to approximately 34 per cent of Mongolia's total GDP (Mining-technology 2011).

Not only is there uncertainty regarding the socio-economic and ecological consequences of Mongolia's unprecedented resource expansion, but projections of permanent changes to Mongolia's already sensitive climate further raise concerns for its long-term sustainable development. Mongolia is recognized as one of the world's most arid regions with desertification affecting '44% to 90% of the country's territory' (Reeves 2011, p. 455). In addition, significant inter- and intra-annual and daily variation in temperature and extremely low rates of precipitation have not only led to intense summer droughts but also a series of harsh winters, commonly known as *dzuds*, causing 'mass debilitation, starvation, and death of livestock that seriously damages the livelihoods of the herder households who depend upon them' (Tachiiri et al. 2008, p. 2252).

Although historically, Mongolians have been successfully adapting to varying climatic conditions, the frequency and intensity of some of these recent events has clearly outstretched existing adaptive capacities, thereby generating significant concerns at both local and national levels. In recent memory, extreme winter dzuds between 1999 and 2003 impacted more than two-thirds of the country's 21 provinces (Batima et al. 2005). With pastoralism as the primary occupation of almost half of Mongolia's total population, damage to livestock and resulting disruption to social-economic wellbeing from these weather events has been beyond repair. More than a quarter of total national livestock perished as a combined effect of harsh living conditions and unavailability of forage following extended periods of summer droughts, thereby leading to a serious socio-economic and ecological crisis in rural remote Mongolia (Batima et al. 2005; Tachiiri et al. 2008).

The simultaneous impacts of a booming resource industry and harsh climatic conditions are most strongly manifested in the deteriorating nature and scope of Mongolia's traditional occupation of livestock herding. Besides several biodiversity and economic impacts, increasing institutional focus on resource expansion has contributed to growing reports of permanent, often irreversible changes to Mongolia's social and cultural identity by forcing, and challenging families that have been herders for generations to '[adapt] a traditional nomadic lifestyle to modern realities' (Sternberg 2008, p. 1294). Mongolia's rapid urbanization (as of 2009, more than 60 per cent of total population in Mongolia was urban) is a sad by-product of increasing pressures of globalization. Currently almost half of the country's rural to urban migrants arrive in Ulaanbaatar, Mongolia's capital city, to live in 'slum-like' conditions with poor infrastructure, and one of the world's highest levels of air pollution (Horwood 2007).

These arguments confirm Mongolia's continuing struggle with a series of complex socio-economic and bio-physical drivers of change that challenge the fundamental fabric of traditional Mongolian society. In light of the popular belief about the country '[standing] at the cusp of a [resource development] process that could have a transformative effect on the economy and prosperity of the state' (Wachman 2010, p. 594), the immediate task for Mongolia's policymakers is to build and support the mining industry as a positive medium that is able to transform its current challenges into economically viable opportunities on the one hand, and provide an institutional climate that sustains the national social and ecological integrity on the other hand.

Finally, growing demand for minerals and resources in the neighbouring countries of China and Russia have provided Mongolia with a hope for economic and subsequent social development over the coming decades. Although public opinion remains divided on whether or not a minerals boom will help 'trickle down' wealth to its impoverished communities or only further the already fast growing gap between Mongolia's rich and poor, the country's resource development experience is still too under-developed to help draw any conclusive observations (Isakova et al. 2012). National and international advocates for Mongolia's future sustainability reiterate the importance of good governance, long-term institutional reforms and large-scale infrastructure investment to help its leaders formulate effective strategies to best capitalize on the socio-economic promise offered by an extraordinary mining boom (Isakova et al. 2012).

Past Experiences of Resource Development with Weak Institutions, Poor Governance and Human Rights Abuse: Implications for Asia's Future

Mining is an economic activity but one with far-reaching social, cultural, ecological and institutional consequences (Ali and Grewal 2006). Whether monetary returns from resource development remain with a select elite group, mostly comprising bureaucrats and business leaders, or are shared more equitably with the larger populace is largely influenced by the nature and quality of institutions and governance structures that are in place to regulate and monitor the resource development process (Mehlum et al. 2006; Wagner and Wellmer 2009). Through regulations and legislation, governments exert institutional force that, in turn, plays an important role in regulating corporate activities and development outcomes (Harrison 2002).

Effective governance is central to not only dealing with negative externalities associated with mining but also maintaining, and where possible, increasing the socio-economic and ecological rate of return from minerals extraction (Waye et al. 2009). A common observation in developing regions is the trend of minerals wealth making the rich richer, often via rampant institutional corruption and continuing neglect of fundamental human rights for a large majority of affected peoples (Leite and Weidmann 1999; Hilson and Maconachie 2009). The resulting resentment and frustration among the disadvantaged, manifested in deepening poverty and conflict, lead to serious repercussions both for local communities as well as for governments and companies operating in these sensitive environments. As Slack (Slack 2009, p. 83) argues, 'Indeed, almost every major new or existing [minerals exploration] project in countries such as Peru, Guatemala, Ghana, Indonesia, and the Philippines is subject at some point to some kind of community protest or expression of concern'.

For the purposes of this chapter, we use the OECD's definition of weak institutions which are characterized by 'environments in which public sector actors are unable or unwilling to assume their roles and responsibilities in protecting rights . . ., providing basic public services . . . and ensuring that public sector management is efficient and effective' (OECD 2006, p. 42). Often, such institutions further lead to 'failures in political, economic and civic institutions' manifested in widespread corruption, low indicators of human development, inability to monitor the activities of select interest groups, and poor international recognition of national and regional public sector management and performance (OECD 2006; Crowson 2009b; McPhail 2009).

These observations have particular relevance for Asia's developing economies which already struggle with hostile policy settings, rapid environmental and social change, and widespread opposition to minerals development, as reported recently in several parts of Indonesia and Papua New Guinea (Ensor 2012; The Jakarta Post 2012). Below we discuss examples drawn together from two of Asia's key mining regions that illustrate how weak policy structures and parochial institutional decisions have led to a range of development conflicts, revoking of mining permits and in some cases, even loss of life.

Mining in Indonesia

As mentioned earlier in the chapter, Indonesia is one of Asia's most prolific mining-intensive countries with minerals contributing significantly to the national GDP. Large economic contributions from resource development have played a crucial role in developing some of its remotest regions such as Papua, West Nusa Tenggara and East Kalimantan (PricewaterhouseCoopers 2011). From a governance standpoint, Indonesia's resource development process is particularly noteworthy in that although the central government is responsible for overarching security and sovereignty with regards to minerals development at the national level, the national decentralization programme in 2000 has since yielded much greater 'power and patronage' to the provincial and local governments in terms of issuing exploration permits and mining licenses (Ballard 2001). As a result, more than 80 per cent of all mining revenues stay with the provincial governments. This has resulted in uncontrolled corruption associated with inexperience and 'the transfer of fiscal and political authority away from the centre . . . [thus creating] new [local institutions as] sites of power struggles over resource management and revenue generation' (Sahin et al. 2012, p. 295). Indonesia's resource development journey is marred with controversial issue of mining permits, lack of accountability on revenue spending, leading to unequal sharing of minerals wealth, and provincial government-led relaxation of environmental regulations to expedite the issue of exploration permits, much to the detriment of local community and ecological wellbeing (Ballard 2001; Sahin et al. 2012, p. 297).

The central versus provincial/local government discord over mining licenses and revenue usage in Indonesia is further compounded by growing community resentment towards mining operations and local government officials (Wahju n.d.). Freeport's Grasberg operation in West Papua is a case in point (Ballard 2001). Once considered Indonesia's 'national asset', the operation has faced – for more than 30 years – ongoing disputes with local communities over poor returns to the local region, loss of livelihoods due to forced resettlements, human rights abuse, and widespread environmental degradation, thereby permanently impacting relations between the community and company (Kyriakakis 2005). More recently, community protests over total lack of consultation with those directly impacted by mining-related exploitation of water supplies and natural forest reserves have resulted in suspending and/or revoking of exploration permits of at least three mining operations – PT Sorikmas Mining in North Sumatra, Australian Arc Exploration in Bima Regency and manganese mining in the West and East Nusa Tenggara regions (Ensor 2012; Sahin et al. 2012; The Jakarta Post 2012).

These examples illustrate how poor planning, manifested in widespread corruption, has led to the accrual of mining benefits only by 'the extractive industries, local elites and government officials while damaging community resources (forests, water supplies) and exploiting community members as a source of cheap labour' (Sahin et al. 2012, p. 303). Although the process of decentralization is widely considered an effective tool to encourage local government officials to embrace future-orientated planning and development for their regions (McPhail 2009), relegating powers without adequate capacity building at the local level to manage high-revenue mining operations has clearly backfired in Indonesia's case. So far, local and provincial authorities in Indonesia have considered mining companies as 'cash cows' and have paid little interest or regard to the welfare of the affected local environment and communities (Ballard 2001). Absence of rigorous social and environmental impact assessments, institutional failure to address the overlapping 'field of power' across different levels of government and the absence of transparency and accountability in decision-making regarding minerals use, exploitation and wealth dissemination have led to numerous conflicts among and between various mining stakeholders including companies, communities and different tiers within the government (Sahin et al. 2012).

Resource development experience in the Philippines

Mining in the Philippines has seen a rapid decline since the 1990s (Stark et al. 2006). From accounting for approximately 20 per cent of total national exports in the 1980s, mining's share fell to less than 2 per cent by the early 2000s (Symon 2007). Several factors are considered responsible for this steep decline, namely: falling commodity prices, political instability, inefficient resource development process, a series of natural disasters and accidents, weak institutional support to the industry, and poor regulatory environment, ultimately leading to public outcry over deteriorating social and environmental quality of life in mine-dependent regions and communities (The Wallace 2003; Symon 2007)

While the national government's 'mining-led development paradigm' continues to seek foreign investment for resource development across the Philippines, there remains strong local opposition to mining generally, particularly among the vast majority of the country's indigenous peoples who inhabit areas rich in mineral resources (Minter et al. 2012). To a large extent, public mistrust 'stems from the visibly sad shape of areas that used to be bustling mine sites before' (The Wallace 2003, p. 60). Widespread pollution of river systems, overflows of tailing storages, and mercury-poisoned coastal waterways causing rapid depletion of fish stocks have significantly impacted public health and wellbeing. Further, destruction of fertile cropping areas and forced resettlement of a large number of indigenous communities not only impacted livelihoods but are also entrenched in public memory, signs of mining legacies that have been addressed neither by local or national institutions, nor by the companies responsible for such uncontrolled destruction. In fact, the current scenario in the country with regards to minerals development is such that 'tensions between government desire to foster the extractive industries and often emotional public suspicion and oppositions are perhaps more pronounced in the Philippines than anywhere else in the [South East Asian] region' (Symon 2007, p. 86).

A critical analysis of the political economy and the history of resource development in the Philippines suggests that several of these negative experiences are further compounded by lackadaisical institutional support and opposing political agendas (Minter et al. 2012). A case in point is the National Commission on Indigenous Peoples (NCIP) in the Philippines. Despite being a government agency responsible for advocating indigenous peoples' rights and facilitating their dialogue with mining companies, NCIP has failed to meet its objectives largely due to its 'lack of independence from a political agenda that is strongly pro-mining' and its inability to generate programmes that consider local socio-economic and political conditions and priorities (Minter et al. 2012, p. 12–13).

These Indonesian and Philippine mining experiences demonstrate that resource extraction across Asia is, to a significant extent, influenced and/or shaped by the political economies, socio-economic evolution and governance ideologies of regions hosting mining activities. As discussed in the next section, many of the challenges highlighted above may be best addressed by developing long-term strategies founded on a set of fundamental elements discussed in the section 'Sustainable Resource Development: Principles and Practice' above. These may include, *inter alia*, a transparent and accountable system of resource management, a diverse export base, a coordinated and timely action plan for mine closures, and an overarching commitment of institutions to harness the socio-economic potential of mineral resources by strengthening partnerships amongst national and local governments, local communities and the resource industry counterparts.

MINING FOR DEVELOPMENT: A PROCESS OF LEARNING, SHARING AND KNOWLEDGE BUILDING

From the perspective of developing countries, despite the range of challenges that accompany resource development, it is widely acknowledged that mining provides several opportunities that wouldn't be available otherwise (Davis and Tilton 2005). As suggested in a report from the United Nations Conference on Sustainable Development (UNCSD) (Australian Government 2011, p. 8), 'when managed properly, mining offers the opportunity to catalyze broad-based economic development, reduce poverty and assist countries in meeting their Millennium Development Goals'. In the past, resources have contributed significantly to shaping the economies of countries such as Chile, Norway, Australia, Canada, Botswana, and the Netherlands. Lessons drawn together from resource development experiences of these countries can provide useful insights for several Asian regions which either currently host mining or plan to focus on resource development for their future socio-economic development.

Policy and Action Strategies for Sustainable Resource Development: Learning the Lessons from Other Regional Experiences

In light of the enormous regional diversity in Asia, it is vital to acknowledge that while a blueprint guide to responsible mining across Asia may be both unrealistic and impractical, drawing together mining experiences from different world regions may better inform Asia's policymakers in addressing potential resource development challenges in the Asian century. In other words, a systematic review of experiences and lessons based on resource development processes undertaken in other parts of the world can help Asia's mineral-rich nations to pre-emptively devise strategies that balance mining-related externalities, while also expediting development goals in a sustainable manner. The list below, while not being exhaustive, provides a wealth of strategies that have proved successful in achieving sustainable resource development in other regions:

1. The establishment of wealth funds in countries such as Chile and Norway to ensure that the profits of resource development collected in the present are not only effectively managed to be available for use by future generations but act as stabilization funds to keep the economy afloat in times of turbulent economic crisis due to a potential fall in commodity prices (Browne et al. 2011);
2. An organized structure of regulatory and legislative demands on resource companies to plan and develop strategies upfront for: consultation and compensation for landholders, including indigenous communities, before proceeding with the project (as practiced in Cape York, northern Australia (O'Faircheallaigh 2009), and New Caledonia (Ali and Grewal 2006); employment, land use, and regional socio-economic development (as practiced in South Africa) (Browne et al. 2011);
3. The establishment of the Responsible Mining Initiative in Mongolia which helped generate a common platform for all actors with a stake in the country's minerals development – including government, non-government organizations (NGOs), industry and community groups – to share their concerns, and devise a consensual plan of action to harness the full potential of minerals development for Mongolia's long-term sustainable development (Browne et al. 2011);
4. The development of innovative industries to help maintain a diversified economic base and limit exclusive dependence on mining revenues for socio-economic development (Slack 2009). In the case of Chile, some of the country's most notable non-mining exports include salmon farming, and wine and fruit production (Davis and Tilton 2005; Havro and Santiso 2011); and
5. The setting up of an institutional structure whereby civil service responsible for devising resource policy is insulated from political influence, to keep checks on corruption and similar anti-development vested interests (Minter et al. 2012). Both Chile and Norway have been extremely successful in ensuring a clear separation of power between policy and politics, supported by cooperative political parties and a favourable policymaking climate in general (Havro and Santiso 2011).

Figure 10.4 further analyses successful mining experiences in other regions to suggest key policy lessons and action strategies that may

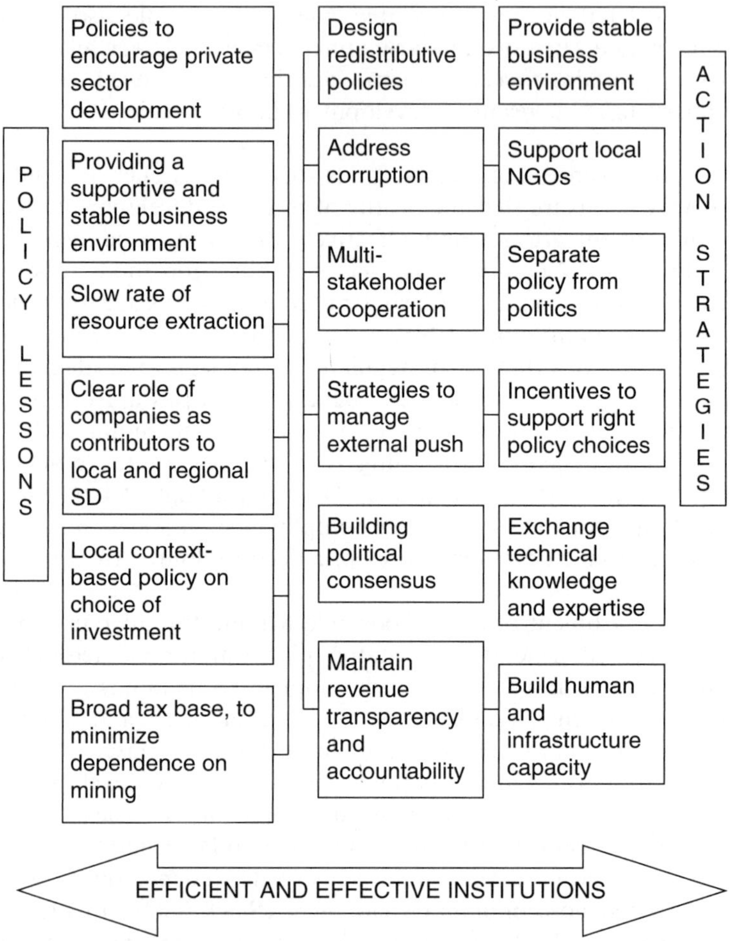

Sources: Davis and Tilton (2005); Symon (2007); Browne et al. (2011); Havro and Santiso (2011); Spiegel (2012); The Jakarta Post (2012).

Figure 10.4 Key policy and strategic lessons for sustainable resource development – based on experiences of other resource regions

strengthen the institutional capacity of resource-rich Asia to transition to a sustainable resource development climate over the next several decades. Based on local socio-political and economic conditions, some of these principles may not be applicable at all, or may require context-driven adaptations. An important element underlying all of these principles is the presence of 'good quality, honest and efficient institutions' (Havro and

Santiso 2011, p. 35). The latter, it is argued in the literature, has the power to shape how the remaining strategies interact with one another to generate positive development outcomes for resource-rich regions (Richards 2009).

Coming Full Circle: Sustainable Resource Development and the Asian Century

So far the discussion above has established that not only does Asia maintain large endowments of mineral resources but that the prospects of converting this resource wealth into long-term sustainable development outcomes rely – quite extensively – on effective and efficient institutional arrangements that are able to generate synergistic cooperation between and among sub-national and the central governments, local communities, and the private sector. In other words, the degree to which resource wealth shapes and influences the future of the Asian century will be determined by the ongoing viability, objectivity and robustness of Asia's institutions and extended relationships they build both domestically and internationally.

Resource extraction provides direct incentives to foster development across several capitals – built/infrastructure, social, human and financial. To this extent, resource wealth, if properly regulated, managed and redistributed, has the potential to set forward a path for holistic development across Asia's many developing regions. Yet, several immediate issues remain which threaten not only the viability of the resources sector itself but also the nature and scope of positive contributions from mineral development to all stakeholders including dependent communities. Below is a list of suggestions whose effective operationalization will ensure that sustainable resource development significantly influences, and further strengthens the future of the Asian century:

- Asia's process of sustainable resource development calls for the design and implementation of strategies that integrate resource extraction with the principles of sustainable development. By considering resources as a positive and indispensable means to development, the demand for better public accounting and reporting by responsible institutions is strengthened;
- It is essential to develop a suite of comprehensive sectoral and institutional level strategies to enable this integration and to successfully make a transition to future-orientated holistic development;
- Sectoral strategies may include: greater emphasis on social, cultural and environmental sustainability; sustainability monitoring,

assessment and reporting; stakeholder participation including formal and informal civic engagement; recognition of local and indigenous knowledge through all stages of resource development, including project planning and decision-making; and embedding sustainability and socio-cultural considerations into decision-making within the government and private sector apparatuses;
- Institutional responses may take the form of establishing effective partnerships among the government, industry and community/ NGOs; greater private-public engagement; and developing governance strategies that foster inclusiveness, transparency and accountability in institutional dealings associated with resource development; and
- Finally, it is important to create and sustain an ongoing climate of governance that supports companies in designing and implementing resource development policies that align globally with the principles of sustainable development as outlined by the ICMM while also being culturally sensitive to the demands, aspirations and priorities of local communities. Such a climate is indispensable to upholding a resource company's 'social license to operate', thereby minimizing conflicts, legal crises and resulting loss of valuable human and financial costs.

The sectoral and institutional strategies referred to above will provide a twofold advantage in managing a positive contribution of resources to sustainable development. Firstly, they will provide a systematic assessment of the 'how' issue at the heart of mining-fuelled growth and development. These strategies can hopefully clarify the role and the nature of the process needed to help integrate resource development with a broader sustainable development agenda at the national and/or sub-regional scale. Secondly, a well-planned array of institutional strategies can contribute to resolving conflicts involving land and water management and use, flow of socio-economic benefits, and ecological integrity between and among different mining stakeholders.

CONCLUDING REMARKS

As a long-term socio-economic activity, mining is paradoxical in nature, particularly for under-developed regions and communities. On the one hand, resource extraction necessitates change across social, economic, environmental and institutional issues. It threatens biodiversity, in some cases causing irreparable damage; causes loss of cultural heritage; exacerbates the rich-poor gap in the absence of effective institutions; creates an unsupportive environment for non-mining traditional economic activities, such as agriculture and livestock herding; and for mining-intensive regions, there is an increasing tendency to remain socially and economically dependent upon mining, thereby leading to dysfunctional governments and societies when mines close. On the other hand, regional cooperation, effective partnerships between different stakeholders, and greater transparency and accountability at the institutional level can address several of these challenges, thus contributing to sustainable resource development.

Owing to vast cultural and socio-economic diversity within Asia, it is virtually impossible to suggest principles of resource development that can be implemented to maintain holistic sustainability across several of its resource-rich regions. However, through a critical analysis of other resource-rich world regions and drawing together their experiences, barriers, and motivations in achieving sustainable resource development, this chapter has made several observations – across both policy and action domains – that may provide useful pointers towards translating Asia's vast natural resource endowments into a source of long-term growth and holistic development. Two of these observations are particularly significant, and encompass: (a) the role of transdisciplinary governance arrangements and institutional innovation as a particularly important element in developing a process through which sustainable resource extraction and development may positively shape the future of the Asian century; and (b) the importance of locally relevant resource development policies that situate resource extraction as a means to generate broader sustainable development outcomes.

In summary, the following quote from Stark et al. (2006, p. 13) aptly summarizes the role of mining in long-term sustainable development:

> Mining is not an economic 'silver bullet' but a potentially promising economic sector that should be viewed in the context of the broader development goals of local communities, provinces, and the nation. Responsible mining requires taking into account not only economic and social effects on the host communities during the life of the mining operation but also the long-term impact of mining activities on those communities in the years after mine closure.

It is hoped that this informed outlook towards minerals development – reiterated throughout the chapter – will provide useful guidance with regards to planning for and operationalizing a smooth transition to sustainable minerals development in Asia over the coming decades.

REFERENCES

Ali, S.H. and A.S. Grewal (2006), 'The ecology and economy of indigenous resistance: divergent perspectives on mining in New Caledonia', *The Contemporary Pacific*, **18** (2), 361–92.

Aubynn, E.A. (2003), 'Community perceptions of mining: an experience from Western Ghana', Department of Earth and Atmospheric Sciences, Edmonton, University of Alberta, unpublished MSc thesis, p. 175.

Australian Government (2011), 'Australia's submission to the Rio+20 compilation document', http://www.uncsd2012.org/rio20/index.php?page=view&type=510&nr=692&menu=20, accessed 2 April 2012.

Ballard, C. (2001), 'Human rights and the mining sector in Indonesia: a baseline study', International Institute for Environment and Development and World Business Council for Sustainable Development.

Ballard, C. and G. Banks (2003), 'Resource wars: the anthropology of mining', *Annual Review of Anthropology*, **32**, 287–313.

Batima, P., L. Natsagdorj, P. Gombluudev and B. Erdenetsetseg (2005), 'Observed climate change in Mongolia', AIACC Working Paper No. 12.

Black, L.D. (2010), 'Measuring the Social License to Operate', Social License to Operate Conference, Brisbane, 8 November.

Brereton, D. and P. Forbes (2004), 'Monitoring the impact of mining on local communities: a Hunter Valley case study', Inaugural Sustainable Development Conference, Minerals Council of Australia, Melbourne.

Brereton, D., C. Moran, G. McIlwain, J. McIntosh and K. Parkinson (2008), *Assessing the Cumulative Impacts of Mining on Regional Communities: An Exploratory Study of Coal Mining in the Muswellbrook Area of New South Wales*, Australian Coal Association Research Program, Brisbane: The University of Queensland.

Browne, W. (n.d.), *Mining in Indochina: The Next Big Dig*, Brisbane: Centre for Social Responsibility in Mining.

Browne, W., D. Franks and G. Kendall (2011), *The Foundations for Responsible Mining in Cambodia – Suggested Approaches*, Phnom Penh, Cambodia: United Nations Development Programme.

Carbonniery, G., N. Wagnerz and F. Bruggerx (2011), *Oil, Gas and Minerals: The Impact of Resource-dependence and Governance on Sustainable Development*, Geneva: Graduate Institute of International and Development Studies.

Cronin, R. and A. Pandya (2009), *Exploiting Natural Resources: Growth, Instability, and Conflict in the Middle East and Asia*, Washington, DC: The Henry L. Stimson Center.

Crowson, P. (2009a), 'Adding public value: the limits of corporate responsibility', *Resources Policy*, **34**, 105–11.

Crowson, P. (2009b), 'The resource curse: a modern myth?', in J.P. Richards (ed.), *Mining, Society and a Sustainable World*, London: Springer, pp. 3–36.

Davis, G.A. and J.E. Tilton (2005), 'The resource curse', *Natural Resources Forum*, **29**, 233–42.

DHVP Research (2011), 'Mining and exploration industries contribute 10 to 11 percent of GDP', 16 April, http://www.vietnamica.net/mining-and-exploitation-industry-contribute-10-to-11-percent-gdp/, accessed 27 March 2012.

DRET (2011), *A Guide to Leading Practice Sustainable Development in Mining: Leading Practice Sustainable Development Program for the Mining Industry*,

Canberra: Department of Resources, Energy and Tourism, Australian Government.

Ensor, J. (2012), 'The perils of doing business without community consent', 3 February, www.abc.net.au/unleashed/3808358.html, accessed 28 March 2012.

Esteves, A.M. (2008), 'Mining and social development: refocussing community investment using multi-criteria decision analysis', *Resources Policy*, **33**, 39–47.

Fast Market Research (10 July 2010), 'Indonesia mining report Q3 2010: new research report', available at Fast Market Research, http://www.prlog.org/10784584-indonesia-mining-report-q3-2010-new-research-report-available-at-fast-market-research.html, accessed 28 March 2012.

Fish, M.S. (2001), 'The inner Asian anomaly: Mongolia's democratization in comparative perspective', *Communist and Post-Communist Studies*, **34**, 323–38.

FlorCruz, J. (2011), 'Mining fuels Mongolia's "wolf economy"', 20 May, http://edition.cnn.com/2011/WORLD/asiapcf/05/19/mongolia.mining.economy/index.html, accessed 27 March 2012.

Franks, D., D. Boger, C. Cote and D. Mulligan (2011), 'Sustainable development principles for the disposal of mining and mineral processing wastes', *Resources Policy*, **36** (2), 114–22.

Gibson, G. and K. Klinck (2005), 'Canada's resilient north: the impact of mining on Aboriginal communities', *Pimatisiwin: A Journal of Aboriginal and Indigenous Community Health*, **3** (1), 116–40.

GSO (2012), 'Employed population at 15 years of age and above as of annual 1 July by industry', http://www.gso.gov.vn/default_en.aspx?tabid=467&idmid=3&ItemID=11673, accessed 28 March 2012.

Guerin, T.F. (2009), 'Improving environmental performance in the minerals supply chain using a life-cycle approach: the role of fuel and lubricant suppliers in enabling sustainable development', in J.P. Richards (ed.), *Mining, Society and a Sustainable World*, London: Springer, pp. 205–50.

Harrison, K. (2002), 'Ideas and environment standard-settings: a comparative study of regulation of the pulp and paper industry', *Governance: An International Journal of Policy, Administration, and Institutions,* **15** (1), 65–96.

Havro, G. and J. Santiso (2011), 'Benefiting the resource rich: how can international development policy help tame the resource curse?', IDS Working Paper 355, Brighton, UK: Institute of Development Studies.

Hill, C. and K. Newell (2009), *Women, Communities and Mining: The Gender Impacts of Mining and the Role of Gender Impact Assessment*, Carlton, Victoria: Oxfam Australia.

Hilson, G. and R. Maconachie (2009), 'The extractive industries transparency initiative; panacea or white elephant for sub-saharan Africa?', in J.P. Richards (ed.), *Mining, Society and a Sustainable World*, London: Springer, pp. 469–91.

Horwood, C. (2007), *Tomorrow's Crises Today: the humanitarian impact of urbanization*, Nairobi: OCHA/IRIN and United Nations Human Settlements Programme (UN-HABITAT).

ICMM (2011), 'ICMM sustainable development framework: 10 principles', http://www.icmm.com/our-work/sustainable-development-framework/10-principles, accessed 27 March 2012.

Indoterra (2011), 'The Indonesian market', 8 February, http://www.indoterra.com/indoterra-market/roktabs.html, accessed 29 March 2012.

Isakova, A., A. Plekhanov and J. Zettelmeyer (2012), 'Managing Mongolia's

resource boom', Working Paper No. 138, London: European Bank for Reconstruction and Development.
Katona, J. (2002), 'Mining uranium and indigenous Australians: the fight for Jabiluka', in G.E.J. and N. Lansbury (eds), *Moving Mountains: Communities Confront Mining and Globalisation*, Goodman, London: Zed Books, pp. 195–206.
Kemp, D. (2009), 'Mining and community development: problems of local-level practice', *Community Development Journal*, **45** (2), 198–218.
Kitula, A. (2006), 'The environmental and socio-economic impacts of mining on local livelihoods in Tanzania: a case study of Geita district', *Journal of Cleaner Production*, **14**, 405–14.
Kyriakakis, J. (2005), *Freeport in West Papua: Bringing Corporations To Account for the International Human Rights Abuses under Australian Criminal and Tort Law*, Melbourne: Faculty of Law, Monash University.
Lawrie, M., M. Tonts and P. Plummer (2011), 'Boomtowns, resource dependence and socio-economic well-being', *Australian Geographer*, **42** (2), 139–64.
Leite, C. and J. Weidmann (1999), 'Does mother nature corrupt? Natural resources, corruption, and economic growth', IMF Working Paper, Washington, DC: International Monetary Fund.
Lyday, T.Q. (2002), 'The mineral industry of the Philippines', http://minerals.usgs.gov/minerals/pubs/country/2002/rpmyb02.pdf, accessed 2 April 2012.
Martens, P. (2006), 'Sustainability: science or fiction?', *Sustainability: Science, Practice, & Policy*, **2** (1), 36–41.
McMohan, G. and F. Remy (2001), 'Key observations and recommendations: a synthesis of case studies', in G. McMohan and F. Remy (eds), *Large Mines and the Community: Socioeconomic and Environmental Effects in Latin America, Canada, and Spain*, Ottawa: International Development Research Centre and The World Bank, pp. 1–38.
McNeill, D., I. Nesheim and F. Brouwer (eds) (2012), *Land Use Policies for Sustainable Development: Exploring Integrated Assessment Approaches*, Cheltenham, UK and Northampton, MA, USA: Edward Elgar.
McPhail, K. (2009), 'The challenge of mineral wealth: using resource endowments to foster sustainable development', in J.P. Richards (eds), *Mining, Society and a Sustainable World*, London: Springer, pp. 61–74.
Mehlum, H., K. Moene and R. Torvik (2006), 'Institutions and the resource curse', *The Economic Journal*, **116** (January), 1–20.
Mining-technology (2011), 'Oyu Tolgoi gold and copper project, Mongolia', http://www.mining-technology.com/projects/oyu-tolgio/, accessed 28 March 2012.
Mining Communications (2008), 'Asia's golden opportunity', February, http://www.mining-journal.com/__data/assets/supplement_file_attachment/0018/123345/AFMAscr.pdf, accessed 1 April 2012.
Minter, T., V. de Brabander, J. van der Ploeg, G.A. Persoon and T. Sunderland (2012), 'Whose consent? Hunter-gatherers and extractive industries in the Northeastern Philippines', *Society and Natural Resources*, DOI: 10.1080/08941920-2012.676160.
O'Faircheallaigh, C. (2009), 'Public policy processes and sustainability in the minerals and energy industries', in J.P. Richards (eds), *Mining, Society and a Sustainable World*, London: Springer, pp. 437–68.
OECD (2002), 'Indonesian Energy and Mineral Resources Development and Its Environmental Management to Support Sustainable National Economic

Development', Conference on FDI and the Environment: Lessons to be Learned from the Mining Sector, OECD Headquarters, Paris, 7–8 February.
OECD (2006), *OECD Risk Awareness Tool for Multinational Enterprises in Weak Governance Zones*, Glossary, Paris: Organisation for Economic Co-operation and Development.
Oum, R. (2009), 'Mining – Cambodia', *Sector Report*, Phnom Penh: UK Trade and Investment.
Owen, D.P. (2007), *Beyond Corporate Social Responsibility: The Scope for Corporate Investment in Community Driven Development*, Washington, DC: The World Bank.
PricewaterhouseCoopers (2011), *Mining in Indonesia: Investment and Taxation Guide*, Indonesia: PwC.
Purba, I.H. (2008), 'Mining sector profile – Jakarta, Indonesia', February, http://www.southernarcminerals.com/_resources/fac_imsp.pdf, accessed 28 March 2012.
Reeves, J. (2011), 'Mongolia's environmental security: Chinese unconscious power and Ulaanbaatar's state weakness', *Asian Survey*, **51** (3), 453–71.
Richards, J.P. (ed.) (2009), *Mining, Society and a Sustainable World*, London: Springer.
Ritter, R.M. (2001), 'Canada: from fly-in, fly-out to mining metropolis', in G. McMahon and F. Remy (eds), *Large Mines and the Community: socio-economic and environmental effects in Latin America, Canada, and Spain*, International Development Research Centre and The World Bank: Ottawa, pp. 223–62.
Robinson, J.A., R. Torvik and T. Verdier (2006), 'Political foundations of the resource curse', *Journal of Development Economics*, **79**, 447–68.
Rolfe, J., B. Miles, S. Lockie and G. Ivanova (2007), 'Lessons from the social and economic impacts of the mining boom in the Bowen Basin 2004–2006', *Australasian Journal of Regional Studies*, **31** (2), 134–53.
Ross, M. (1999), 'The political economy of the resource curse', *World Politics*, **51** (2), 297–322.
Sachs, J. and A. Warner (2001), 'The curse of natural resources', *European Economic Review*, **45**, 827–38.
Sahin, S., B. Lewis and J. Lewis (2012), 'Fractured futures: Indonesian political reform and West Timorese manganese mining', *Global Change, Peace & Security*, **24** (2), 289–304.
Shepard, R.B. (2008), *Gaining a Social License to Mine*, Troutdale, OR: Applied Ecosystem Services Inc.
Slack, K. (2009), 'The role of mining in the economies of developing countries: time for a new approach', in J.P. Richards (eds), *Mining, Society and a Sustainable World*, London: Springer, pp. 75–90.
Solomon, F., E. Katz and R. Lovel (2008), 'Social dimensions of mining: research, policy and practice challenges for the minerals industry in Australia', *Resources Policy*, **33**, 142–9.
Sothea, R. (2010), 'After long wait, Cambodia opens to mining', *Voice of America*, http://www.voanews.com/khmer-english/news/economy-business/a-40-2010-03-18-voa7-90235907.html, accessed 28 March 2012.
Spiegel, S. (2012), 'Governance institutions, resource rights regimes, and the informal mining sector: regulatory complexities in Indonesia', *World Development*, **40** (1), 189–205.

Stark, J., J. Li and K. Terasawa (2006), 'Environmental safeguards and community benefits in mining: recent lessons from the Philippines', Working Paper No. 1, Washington, DC: Foundation for Environmental Security and Sustainability, USAID.
Sternberg, T. (2008), 'Environmental challenges in Mongolia's dryland pastoral landscape', *Journal of Arid Environments*, **72**, 1294–304.
Symon, A. (2007), 'Petroleum and mining in Southeast Asia: managing the environmental and social impacts', *Southeast Asian Affairs*, 2007, 77–100.
Tachiiri, K., M. Shinoda, B. Klinkenberg and Y. Morinaga (2008), 'Assessing Mongolian snow disaster risk using livestock and satellite data', *Journal of Arid Environments*, **72**, 2251–63.
The Jakarta Post (2012), 'Indonesia: protesters set Bima Regency office on fire', January, http://www.minesandcommunities.org/article.php?a=11459&l=1, accessed 25 March 2012.
The Wallace (2003), 'Philippine mining: it can play a positive role', http://denmar1international.tripod.com/sitebuildercontent/sitebuilderfiles/miningpositiverole.pdf, accessed 2 April 2012.
Thomas, G. and M. Nowak (2006), *Corporate Social Responsibility: A Definition*, Perth, WA: Graduate School of Business, Curtin University of Technology.
Wachman, A.M. (2010), 'Suffering what it must? Mongolia and the power of the "weak"', *Orbis*, Fall, 583–602.
Wagner, M. and F.-W. Wellmer (2009), 'A hierarchy of natural resources with respect to sustainable development – a basis for a natural resources efficiency indicator', in J.P. Richards (ed.), *Mining, Society and a Sustainable World*, London: Springer, pp. 91–121.
Wahju, W.N. (n.d.), 'Decentralizing mineral development efforts: the case of Indonesia', http://www.ima-api.com/downloads.php?pid=42&cat=2, accessed 30 March 2012.
Waye, A., D. Young, J.P. Richards and J.A. Doucet (2009), 'Sustainable development and mining – an exploratory examination of the roles of government and industry', in J.P. Richards (ed.), *Mining, Society and a Sustainable World*, London: Springer, pp. 151–82.
WCED (1987), *Our Common Future*, World Commission on Environment and Development, Oxford University Press.
Wu, J.C. (2001), 'The mineral industry of Vietnam', http://minerals.usgs.gov/minerals/pubs/country/2001/vmmyb01.pdf, accessed 28 March 2012.
Wu, J.C. (2003), 'The mineral industries of Cambodia and Laos', http://minerals.usgs.gov/minerals/pubs/country/2003/cblamyb03.pdf, accessed 2 April 2012.

11. Business's challenge: relating corporate sustainability, strategy and leadership

Vikram P. Murthy

DEFINING THE SUSTAINABILITY CHALLENGE AND EXPLAINING THE RATIONALE FOR FIRM-CENTRIC INQUIRY

This chapter diverges in one significant respect from those that have preceded it – it uses the individual firm or business as its lens to view the sustainability challenge. It is from this vantage point that the 'driving forces' (Wilkinson 2009) in the mega-environment, of 'rapid technological diffusion', 'extensive environmental threats', and 'vast current inequalities of income and power' (Sachs 2008, p. 8) are viewed. The dystopian scenarios of undermined global governance, social unrest, and ineffective safeguards that may arise from these critical uncertainties (WEF 2012, pp. 16–27) are also contemplated from this firm-centric mindset.

The publication of the first edition of the Global Risks Report of the World Economic Forum in 2006 is a temporal identifier for this prevailing era of relentless and dynamic uncertainty that the above description foregrounds. It is conceptually useful because it highlights the accelerating social, environmental and economic issues as the unfortunate but unavoidable externalities of over two centuries of anthropocene to that point, an epoch where humanity has seriously disturbed many critical earth systems (Crutzen 2009). Amongst the more impactful and potentially deleterious markers of business's macro-environment landscape in this era are chronic fiscal imbalances, greenhouse gas emissions, global governance failure, unsustainable population growth, critical systems failure, overexploitation of species, and mismanaged natural cycles (WEF 2012, p. 11; Sachs 2010, pp. 7–9). This is the overarching sustainability challenge.

There are two compelling reasons for making business the centrepiece of any inquiry into responses to this challenge. The first is a present and clearly articulated understanding on the part of government and civil society that, 'business leadership is needed in adopting efficiency measures, mobilising capital, creating new markets, developing new technologies, driving innovation, deepening our skills base and developing partnerships across the whole community' (Rudd 2008). The second is the complementary realization on the part of business that it must pro-actively lead on this issue because by 'playing the bystander' it will 'recklessly' ignore 'a significant shift in its market environment' (Hoffman 2007, p. 34).

SUSTAINABILITY AND BUSINESS'S NEW REALITIES

The consequent conundrum for business of this shift in its environment is best described in Lubin and Esty's (2010) summation:

> Environmental issues have steadily encroached on businesses' capacity to create value for customers, shareholders, and other stakeholders. Globalised workforces and supply chains have created environmental pressures and attendant business liabilities . . . Externalities such as carbon dioxide emissions and water use are fast becoming material. (p. 44)

Business is struggling with the paradox of redirecting its attention from its single traditional objective of financial performance to two additional vectors of strategic performance, social and environmental sustainability. When viewed from the purely financial corner of the proposition (McDonough 2009), the apparently conflicted nature of the demands of social and environmental sustainability goals on the one hand and the financial and commercial objectives of business on the other hand, are in sharp relief. It is this counterpoint that Siegel (2009) asserts when he says that: 'Managers of publicly traded firms have a fiduciary responsibility to adopt "green management" practices only if such actions complement the organization's business and corporate-level strategies' (p. 5).

This suggests that business must actively seek to identify, understand and build those new strategic capabilities that will help it succeed financially in a sustainable world. More importantly it must do all this using the 'tools of the strategist' so that the 'issue is best addressed' (Porter and Reinhardt 2007, p. 22). It will now be argued that deploying these tools however requires rooting the sustainability challenge in existing strategy paradigms.

INCORPORATING THE SUSTAINABILITY CHALLENGE INTO BUSINESS STRATEGY

Porter and Kramer (2006) make a compelling argument that business 'must integrate a social perspective into the core frameworks it already uses to understand competition and guide its business strategy' (p. 84). They contend that any sustainability approach that is 'fragmented' and 'disconnected from business and strategy' will 'obscure many of the greatest opportunities for companies to benefit society' (p. 80). The proliferation of overlapping 'concepts such as sustainable development, corporate citizenship, sustainable entrepreneurship, Triple Bottom Line, business ethics, and corporate social responsibility' and the ensuing confusion and impediments in implementation of a coherent corporate sustainability strategy are both an evidence and outcome of such fragmentation (van Marrewijk 2003, pp. 95–96; Faber et al. 2005).

The case therefore exists for integrating sustainability issues into business strategy and more importantly thereafter, using these issues as levers for value-creation and sustainable competitive advantage. Yet, as Gladwin et al. (1995) observed, 'attention to nonhuman nature is absent from the strategic management literature', in the period leading up to 1994 (p. 874).

In the 15 years thereafter to the end of the first decade of the new millennium, the progress on the subject of 'making social impact integral to the overall strategy' (Porter and Kramer 2006, p. 90) has remained desultory as underlined by Crews' (2010) observation that 'the focus of much of the research [in sustainability] is on evaluation and monitoring, with little attention to strategy formulation, and implementation' (p. 15).

THE FIRM'S STRATEGIC CAPABILITIES FOR ADAPTING TO THE SUSTAINABILITY CHALLENGE

The starting point for such integrated strategy formulation is conceptualizing a sustainability framework for the business. As Selznick (1957) advised so presciently more than 50 years ago, this requires understanding the business's 'external expectation' and matching that with its 'internal state' by developing 'distinctive competences' (pp. 67–74). External expectations such as environmental regulation, standards set by governments, and pressures resulting from customer groups and community provide useful stimuli to business (Howard-Grenville 2006). They signal the 'foreseeable change in the social, political, and macroeconomic context' of the firm and its industry (Christensen et al. 1982, pp. 179–80).

Business's response to these signals provides a partial understanding of how external environment and the opportunities and threats therein, guide its actions. From a sustainability perspective, it explains, for example, the forces that drive it to minimize emissions and waste, limit life cycle cost of products, and reduce the environmental burden of firm growth (Hart 1995, p. 992). However if these merely result in the development of practices with the limited objective of control and compliance, then business's response has been partial, reactive and tactical. What it needs to fully comprehend in addition to its external expectation therefore, is the structure and content of its internal state and the nature of its distinctive competence: that is, its key resources and capabilities and the sustainable competitive advantages that could potentially accrue from their leverage for overcoming the sustainability challenge.

SUSTAINABILITY CAPABILITY AS THE SOURCE OF SUSTAINABLE COMPETITIVE ADVANTAGE

One strategy paradigm that recognizes the potential of valuable, rare, inimitable and non-substitutable firm resources to generate competitive advantage is the resource-based view (Barney 1991, p. 99; Barney 2001; Wernerfelt 1984). Augmenting this paradigm in the light of the sustainability challenge Hart (1995) has argued: 'It is likely that strategy and competitive advantage in the coming years will be rooted in capabilities that facilitate environmentally sustainable economic activity – a natural-resource-based view of the firm' (p. 991).

Hart (1995) has proposed a conceptual classification for this natural-resource-based view, where three strategic capabilities of pollution prevention, product stewardship, and sustainable development, have been founded on three key firm resources of continuous improvement, stakeholder integration and shared vision respectively. These three strategic capabilities could each then conceivably yield the business sustainable competitive advantages. Pollution prevention, which seeks to prevent waste and emissions, could give a sustainable competitive advantage of overall lower costs. Product stewardship, which expands this prevention to include the entire value chain and life cycle of the business's product systems, creates the potential for competitive advantage through strategic pre-emption. Sustainable development, which reduces the environmental burden and increases the economic benefits for lesser-developed markets, could deliver a competitive advantage of future position (Hart 1995, p. 992; Hart and Dowell 2011, p. 1466).

Subsequent iterations of the original classification, additions to it from other strategy paradigms, and practitioner and scholarly inputs have made the natural-resource-based-view comprehensive and robust. For example, the strategic capability of sustainable development has been separated into two distinct capabilities: clean technology and Base of Pyramid (BoP) (Prahalad and Hart 2002). In addition, the explicatory strengths of the dynamic capabilities perspective, with its emphasis on adaptation within ambiguous and dynamic markets, have added to the paradigm. It has, for example, informed the study of clean technology and BoP, because the operating context for firms developing the capabilities to deal with these issues resonates with the dynamic capabilities perspective (Teece et al. 1997, p. 509; Prahalad and Hamel 1990; Dierickx and Cool 1989; Aragon-Correa and Sharma 2003).

This amplified and augmented natural-resource-based view resonates with sustainability experts and their 'roadmap for natural capitalism' (Lovins et al. 2007, p. 172). The strategic capabilities that it enumerates correlate well with the four interlinked shifts in business that proponents of natural capitalism advocate: firstly, radical resource productivity through cradle-to-cradle and whole-system design tenets (McDonough et al. 2003, p. 436) and innovative, environmentally friendly technologies; secondly, biomimetic production with closed biological and technical nutrient cycles (Benyus and Baumeister 2002, p. 26; McDonough 2005, p. 36); thirdly, a solutions-based business model where value is delivered as a flow of services (Lovins et al. 2007, p. 174); and lastly, reinvestment in natural capital through an augmented multiple capitals framework (Bent and Draper 2007; Porritt 2006).

This significantly enhanced version of the natural-resource-based view also makes common voice with academics who have been insisting for almost two decades that firms needed to cultivate an emerging set of sustainability related capabilities like waste minimization, green product design, and technology cooperation in the developing world in order to enjoy a sustainable competitive advantage in the future (Gladwin 1992; Kleiner 1991). It is simpatico with the steady inclusion of a growing list of critical capabilities into the firm's repertoire. They include for example: the ability to anticipate and shape regulation; the capability to manage green know-how; and the ability to understand what customers want and to figure out different ways to meet their needs (Nidumolu et al. 2009, p. 60).

THE CAUSAL CHAIN OF SUSTAINABLE COMPETITIVE ADVANTAGE

Potential sustainable competitive advantages can accrue to the business as a consequence of the augmented natural-resource-based view described in

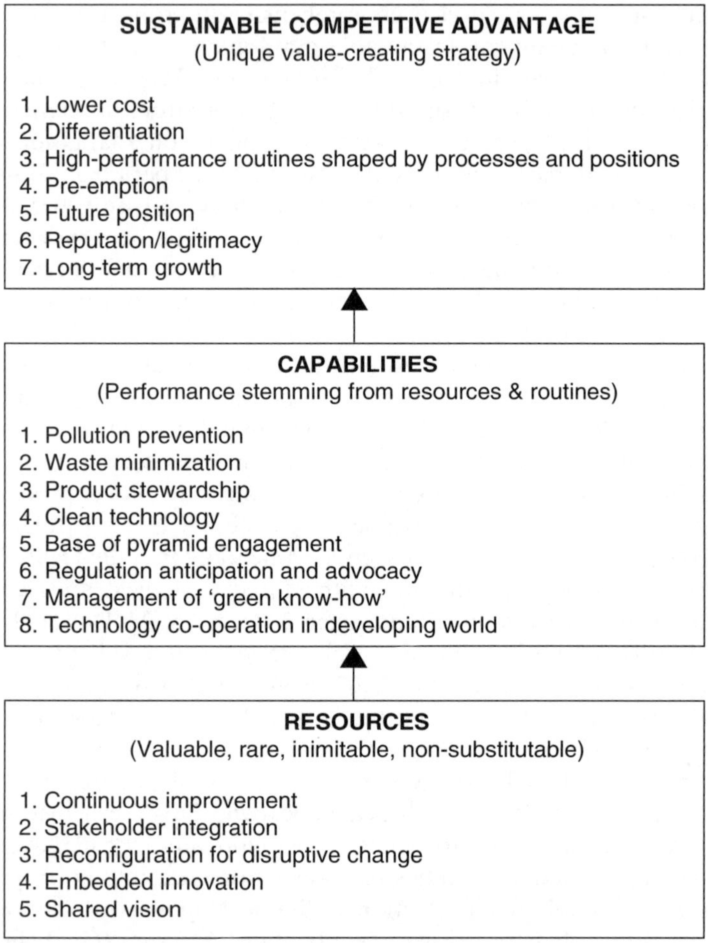

Figure 11.1 The natural environment and firm's resources, capabilities and competitive advantage

the preceding discourse. This is because of the causative flow of strategic capabilities from sustainability related resources (see Figure 11.1).

The need to develop and hone sustainability-related resources and capabilities as the basis of strategic competitive advantage is therefore critical. Absent this capability and 'the natural environment could create a serious constraint on firms' attempts to create sustainable advantage' (Hart and Dowell 2011, p. 1465). The presence of these strategic capabilities on the other hand will help firms to simultaneously improve social and human welfare, reduce their ecological impact and effectively achieve organizational objectives (Sharma 2003).

SIGNIFICANCE OF CORPORATE SUSTAINABILITY AS A STRATEGIC CAPABILITY FOR THE DEVELOPING WORLD

The preceding arguments are of equal import to both developing countries and the developed world. This is because developing countries are bearing the brunt of the impact of phenomena like globalization, foreign direct investment, economic growth and business activity (World Bank 2006). Globalization in particular has evidenced markedly deleterious outcomes for developing countries and its ensuing problems have been increasingly intractable and therefore difficult to ignore.

The responses to the common challenges may however manifest differently in these countries because of specific and contextualized preoccupations. Thus, for example, the social theme is the pressing focus for business's role in society in developing countries as opposed to, for example, business ethics, or stakeholder management. Corporate social responsibility (CSR) is thus synonymous with corporate sustainability in the developing world (Visser 2008, p. 475). This is evidenced in India, for example, where 'there is a strong belief that CSR is an essential element in social uplift and development, something very relevant to India' (Balasubramanian et al. 2005, p. 86).

In turn this has meant that CSR in developing countries is itself yoked to a variety of economic multipliers, including, but not limited to, the capacity to generate investment and income, create jobs, invest in human capital, spread international business standards, produce safe products and services, and build physical and institutional infrastructure (Nelson 2003). This is understandable because if one takes India as an example, even while its corporations have become globally competitive, the country still battles with issues like high unemployment, income inequality, and lower standards of health, education, safe drinking water, nutrition and so on (Arora and Puranik 2004).

BENEFITS OF CSR FOR BUSINESS IN EMERGING ECONOMIES

Business therefore has a pragmatic rationale for such an economic focus, given that it 'is negatively affected by poor education, poor health, delinquency, crime, and unstable regional and national development' (Boyle and Boguslaw 2007, p. 103). Under the circumstances the corporate sector will be a key beneficiary of any systemic solution that can visualize investment in the development and growth of individuals, civil societies and nations as a critical necessity (Karnani 2007).

Businesses in emerging economies apparently understand the above argument, judging by the available empirical evidence on CSR from the developing world. Thus for example in the period just after the millennium, nearly three quarters of large companies in India, the emerging economy exemplar referred in this chapter, presented themselves as having corporate social responsibility policies and practices (Chapple and Moon 2005).

The justification for this has come from turn-of-the-millennium business strategy overhauls that advantaged longer-term business initiatives (Khan and Faisal 2001) and supported social responsibility concerns on the part of business (Sankaran 2003). This salutary state of CSR appears to be continuing and paradoxically for a country with the lowest level of gross national product per capita, India has the highest level of CSR among other Asian economies (Reserve Bank of India 2009).

INTEGRATION INTO CORPORATE STRATEGIES AND DAILY PRACTICE

'Nonetheless', as Baskin (2006) has argued, 'corporate sustainability in emerging markets, while more extensive than commonly believed, is less embedded in corporate strategies, less pervasive and less politically rooted than in most high-income OECD countries' (p. 46). Indian academic, practitioner and institutional commentators concur that while CSR is being seen as a strategic concern for corporations in India, it has not been extensively integrated into the daily practices of many of the large Indian corporations (Balasubramanian et al. 2005, p. 87).

They urge that effort is needed to establish the principles of CSR as part of normal strategy and business operations (NSE/NIFTY 2003; Sankaran 2003). As the experience of transnational corporations has demonstrated, these exhortations may be well placed and worth heeding because, 'these companies are in fact deriving strategic value from responsible business practice . . . even finding that principled behaviour is essential to business survival and success' (Hall 2007, p. 31).

The defining nature of the challenge of sustainable development for both emerging economies and for the developed world is evident from the preceding discussion. Business cannot however embark on strategic corporate sustainability capabilities-building activities while ignoring the role of organizational leadership in effecting such activities. This would be a material omission that may 'yield imprecise conclusions regarding the antecedents and consequences of such activities' (Siegel 2009, p. 13). The remainder of the chapter explores organizational leadership for corporate sustainability.

Emerging Sustainability–Leadership Lessons from Praxis

With the foregoing as the backdrop, a review of the extant practitioner literature for business exemplars of corporate sustainability will help underline insights on leadership virtues, practices and enactments in these organizations. The evidence cited below will argue that the nature of organizational leadership, its pervasiveness and performance are crucial predictors of firm success when responding to the challenge of sustainability. Successful corporate sustainability planning begins with the realization that to remain 'relevant' in a 'Green, Global, Google' world, organizations must transform into 'living, dynamic systems of interconnected relationships ready to change in smart ways' (Duin and Baer 2010, p. 31; Epstein et al. 2010). For such a transformation to be systemic it must be pivoted on conceptions of 'leadership as a more relational process' that is dependent on 'networks of influence' (Fletcher and Kaufer 2003, p. 21).

On the one hand, such transformation must be embedded in an organizational culture that fosters leadership at all levels because as the earlier discussion on social movements has highlighted, 'evenly shared' power ensures 'faster response times to environmental demands' during turbulent times (Pearce et al. 2009, p. 235). On the other hand, it must attract and hold senior leadership's strategic interest in order to gain organizational legitimacy and operational momentum. As Lubin and Esty (2010) have deduced from their consulting experience, senior leadership is indispensable to the success of sustainability initiatives because:

> Companies that excel in sustainability make shifts in key areas, moving from tactical, ad-hoc, and siloed approaches to strategic, systematic and integrated ones . . . [Their] C-level leadership links sustainability strategy, with initiatives and outcomes, and develops shared goals in partnership with suppliers, customers, and other stakeholders. (p. 46)

Case studies from large American businesses like Procter and Gamble, and Nissan North America, buttress these arguments. They provide empirical validation that 'soft systems' like 'organizational culture, leadership and people', do 'nurture a company's drive for sustainability' and 'integrate sustainability into the rhythm' of its business (Epstein et al. 2010, pp. 44–5). In a similar vein, academic research has also begun to underline the significance of 'the role of corporate leaders in formulating and implementing' sustainability initiatives (Siegel 2009, p. 13). Reminiscent of the concept of organizational task complexity and leadership's situational response in contingency theory of leadership, Heifetz and Laurie (1997) have argued that the '*adaptive challenges*' [italics in the original] organizations face because of 'changes in societies, markets, customers, competition, and technology around the globe' demand 'new ways of operating' (p. 124).

Writing more than a decade later, Kegan and Lahey (2010) stress that these adaptive challenges are continually becoming more nuanced in a 'faster, flatter, [and] more interconnected' world (p. 782). 'Leading business' therefore involves 'engaging people in collective goals that damage neither people nor environments' (Hind et al. 2007, p. 8). This implies that for the goals of sustainability to be achieved, leaders must 'reform, redesign, and restructure their organizations to minimise their negative impacts' (Quinn and Dalton 2009, p. 22).

ZEITGEIST LEADERSHIP: INTEGRATING COGNITION, CONSCIENCE AND COLLECTIVE SPIRIT

The preceding examples and arguments for a variety of leadership practices and enactments are being more fully investigated, developed and classified using the ongoing findings of two related research studies that have been undertaken in Australia and New Zealand respectively. Both studies use a neo-classical grounded theory approach (Glaser and Strauss 1967) to explore enhanced leadership virtues, practices and enactments that comprise strongly emerging and conceptually appropriate leadership responses to the challenges and opportunities of a dynamically changing environment.

The first of these studies involved senior leaders of eight successful Australian-based local, regional, and multinational businesses. The outcomes have been detailed in Murthy and McKie's (2009) book *Please Don't Stop the Music: An Ensemble Leadership Repertoire for Productive Sustainability, and Strategic Innovation in Uncertain Times*. Two recurrent themes in Australian organizations wrestling with high-velocity change in 2008, just prior to the Global Financial Crisis, have been described in this book. The first was the environment as a prime driving force in organizations' performance; and the second was its deterministic influence on leadership and management's intentionality and choice. The authors surfaced a set of newly evidenced, and still-developing challenges, which promised to have explicatory power when it came to describing what 'environments of high-velocity change' actually meant in practice, to businesses. They also signalled a set of emerging leadership practices relevant to these high-velocity and complex environments.

The grounded theorizing thus produced a set of theoretical concepts that collectively defined the core category of 'Complex Environment'. It also established that leadership's successful response to this Complex Environment is an augmented leadership repertoire described by another set of theoretical concepts labelled the Ensemble Leadership Repertoire. These two core categories of Complex Environment and Ensemble Leadership Repertoire that emerged from this grounded research informed and extended theoretical understanding in both areas in new and insightful ways.

When it came to the complex environment, it asserted that there is a definitive and dramatic difference in the current environment, making it unique when compared to anything that preceded it. This was strongly evidenced from the fact that each of the three concepts that emerged from the senior leaders' substantive descriptions of the environment, were founded on incidents that were products of post-millennial issues and concerns, not experienced at any time prior. Secondly, it distilled the amorphous and multiple descriptors of the current and likely future environment into three internally consistent, but mutually differentiated concepts. Thirdly it added clear dimensions to each of these three concepts by weighting them based on the relative importance assigned to them in leaders' interviews: with 47 per cent of the complexity arising from the internal environment; 29 per cent coming from the industry in which they participate; and 24 per cent resulting from global driving forces (see Figure 11.2).

The grounded theory research provided a robust and reliable basis for leaders and their organizations to develop practices and frameworks that targeted the environmental complexity detailed in this study. The research study generated an augmented leadership repertoire of existing and emergent practices to successfully navigate the current and future challenges of a Complex Environment. The Ensemble Leadership Repertoire provided a robust and reliable theory for use in a post-millennial world as it was generated from rich data, densely conceptualized and delineated into three clear categories. Each of the three categories was clearly dimensionalized based on the relative importance assigned to them in leaders' interviews. The results highlighted 'Zeitgeist – integrating cognition, conscience and collective spirit' – as a new and emergent category that existing leaders recognized as highly significant based on the relative importance this category was assigned in leaders' interviews – 66 per cent (Figure 11.3).

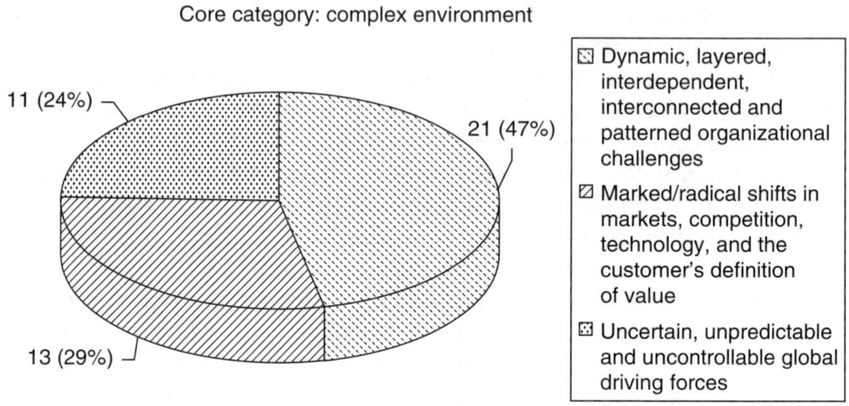

Figure 11.2 The complex environment and its dimensionalized concepts

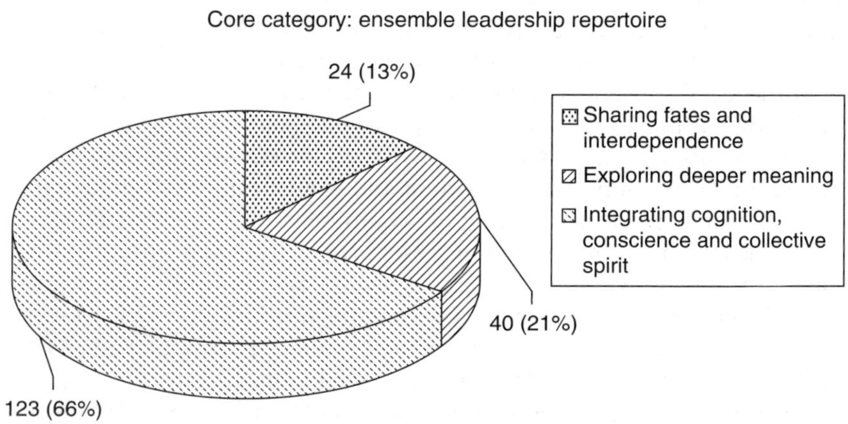

Figure 11.3 The ensemble leadership repertoire and its dimensionalized categories

The Australian research study has been extended to New Zealand where the two themes – of the environment as a prime driving force and leadership's corresponding adaptive repertoire – are continuing to undergird the work of longitudinal exploration with senior organizational leaders. This stage of the study currently involves the senior leadership of 18 multinational, regional, and local organizations across several sectors (Table 11.1).

The study has narrowed its focus to just one emerging category of the full Ensemble Leadership Repertoire inducted from the Australian

Table 11.1 Demographics of NZ study

Industry	Key products & services	Position/role
Charity	Disability related services	Director HR
Construction	Marine, civil, mechanical, maintenance services	Managing director
Construction	Commercial construction	Finance director
Digital media	Digital marketing tools	Managing director
Education	Tertiary education	Vice chancellor
Energy	Electricity distribution services	CEO
Financial services	Banking	General manager
FMCG	Printing & packaging	Managing director
Hospitality	Pubs & restaurants	Managing director
Hospitality	Event management	CEO
Local council	City services	CEO
Manufacturing	Machinery design & manufacture	CEO
Professional association	Business advisory	CEO
Real estate	Property investment & development	CEO
Recruitment	Executive search & placement services	Managing director
Telecommunications	Mobile telecommunications	Chief Marketing Officer
Telecommunications	Mobile telecommunications	General Manager
Telecommunications	Mobile telecommunications	Director HR
Telecommunications	Mobile telecommunications	CEO
Tourism	Marketing, tourism & strategy	CEO
Training & development	Management training	Managing director

research that preceded it. This is the category of Zeitgeist – Integrating Cognition, Conscience and Collective Spirit. Initial findings from the New Zealand study suggest a classification of five virtues', with their corresponding practices and related enactments that could describe this Zeitgeist Leadership response to uncertain times.

The leadership virtue of 'being present' has two key practices: 'excelling at work'; and 'developing self'. Each of these two practices is underpinned by important enactments that operationalize the practice (Table 11.2).

The leadership virtue of 'being good' has three key practices: 'taking a strengths-based orientation to organizational and personal change'; 'leading a full life'; and 'leading ethically'. Each of these three practices is underpinned by important enactments that operationalize the practice (Table 11.3).

Table 11.2 The virtue of 'being present'

Practice: excelling at work Enactments	Practice: developing self Enactments
1. Storytelling	1. Acting authentically
2. Responding wisely & adaptively to the environment	2. Developing self awareness
3. Performing efficaciously	3. Moving forward with energy & purpose
4. Maximizing positive inner work-life	4. Constructing extended life narratives
5. Predicting 'flow' on signature strengths	5. Growing through adversity
6. Searching for enjoyed absorption & meaning	6. Deploying mature defence mechanisms
7. Using skill & effort for achievement	7. Building self through choice of action
8. Demonstrating 'grit' for high accomplishment	

The leadership virtue of 'being in touch' has three key practices: 'making organizational learning transformative'; 'inquiring collaboratively'; and 'understanding the other'. Each of these three practices is underpinned by important enactments that operationalize the practice (Table 11.4).

The leadership virtue of 'being creative' has three key practices: 'managing mindfully during uncertain times'; 'projecting horizons of future expectations'; and 'reinventing buyer value'. Each of these three practices is underpinned by important enactments that operationalize the practice (Table 11.5).

The leadership virtue of 'being global' has three key practices: 'respecting the realities of prevailing times'; 'balancing the paradox of redirecting attention'; and 'leading diversity and multiculturalism'. Each of these three practices is underpinned by important enactments that operationalize the practice (Table 11.6).

CONCLUSION

This chapter uses the individual firm or business as its lens to view the sustainability challenge. It makes the case for integrating sustainability issues into business strategy and more importantly thereafter, using

Table 11.3 The virtue of 'being good'

Practice: taking a strengths-based orientation to organizational & personal change	Practice: leading a full full life	Practice: leading ethically
Enactments	Enactments	Enactments
1. Making people's strengths effective & their weaknesses irrelevant	1. Seeking happiness through personal violation	1. Attending to personal values
2. Inquiring into 'what works well'	2. Learning respectfully from the past	2. Reassuring followers by providing sense & order
3. Encouraging collaboration & collegial effort	3. Viewing the future with hope & optimism	3. Liberating followers' creativity & spirituality
4. Evolving affirmative images	4. Meaning through generativity	4. Finding fulfilment in helping followers reach their potential
5. Being the change you wish to see	5. Connecting individuals' happiness to the organization's success	5. Sharing power & authority
6. Flourishing through positivity		6. Making common cause with followers' vision, tasks & goals
		7. Protecting the welfare of others outside the group

these issues as levers for value-creation and sustainable competitive advantage.

The starting point for such integrated strategy formulation is conceptualizing a sustainability framework for the business. The augmented natural-resource-based paradigm provides this by rooting strategy and competitive advantage in capabilities that facilitate environmentally sustainable economic activity.

Subsequent iterations of the original classification, with additions from other strategy paradigms as well as practitioner and scholarly inputs, have made the enhanced natural-resource-based-view comprehensive and

Table 11.4 The virtue of 'being in touch'

Practice: making organizational learning transformative	Practice: inquiring collaboratively	Practice: understanding the other
Enactments	Enactments	Enactments
1. Fresh thinking & new learning	1. Leveraging collective wisdom	1. Accurately detecting others' emotions for effective deliberate action
2. Adapting to change on an ongoing basis	2. Learning to act & acting to learn	2. Respecting subjective reality
3. Going beyond the problem solving orientation	3. Matching learning styles with learning opportunities	3. Valuing diverse views & perspectives
4. Boundaryless collaboration	4. Reflecting on action not just on planning	4. Improving the ability to empathize through contemplative practice
5. Acknowledging gaps in knowledge & action	5. Mindset of inquiry not just advocacy	5. Communicating openingly
6. Celebrating honourable failure	6. Asking the right questions to spur creative thinking	6. Inquiring into others' thinking & reasoning
7. Harnessing inter-personal conflict for learning	7. Surfacing unasked questions on issues	7. Valuing interdependence
8. Challenging & overcoming organizational defence mechanisms	8. Building group dialogue & cohesiveness through questions	
9. Encouraging individual critical reflection	9. Improving the quality of actions & learning by better questioning	
	10. Challenging the validity of pre-suppositions in prior learning	

Table 11.5 The virtue of 'being creative'

Practice: managing mindfully during uncertain times	Practice: projecting horizons of future expectations	Practice: reinventing buyer value
Enactments	Enactments	Enactments
1. Demonstrating new ways of seeing, thinking & acting	1. Understanding the 'long-fuse, big bang' nature of strategic decision making	1. Orienting the organization to innovation & a willingness to pursue new opportunities
2. Sensing the unexpected in a stable manner	2. Cultivating the long view by using cognitive processes	2. Having an 'outside-in' perspective
3. Dealing with the unexpected in a variable manner	3. Identifying long term driving forces in the environment	3. Targeting value innovation by reconstructing 'buyer value elements'
4. Focusing on the big picture of the future	4. Engaging in the scenario thinking to catalyse bold action	4. Shifting competitive boundaries to create new market space
5. Knowing the detailed picture in the moment	5. Using scenarios to engage with uncertainty systematically	
6. Willingness & capability to invent new expectations		
7. Responding uniquely to specific problems & opportunities		
8. Concentrating on ongoing learning & renewal		
9. Rethinking the organization's structure & hierarchy		
10. Understanding the power of new organizational metaphors to conceptualize change		

Table 11.6 The virtue of 'being global'

Practice: respecting the realities of prevailing times	Practice: balancing the paradox of redirecting attention	Practice: leading diversity & multiculturalism
Enactments	Enactments	Enactments
1. Appreciating that organizational circumstances co-relate with its particular responses	1. Understanding trade-offs in corporate responsibility	1. Recognizing that one person's values maybe different to the next
2. Recognizing emergence of a new world order	2. Recognizing 'business as usual' policies will not lead to success	2. Exploring the deeper implications of cultural artefacts
3. Adopting an emergent approach to management practices	3. Balancing world views with regard to environmental sustainability with economic growth & social development	3. Appreciation of the cultural values behind cultural artefacts
4. Valuing individuals' contribution to organizational innovation	4. Identifying growth opportunities from sustainable development	4. Integrating world views of different cultures
5. Recognizing that strategic performance has been redefined in a globalized world	5. Leveraging the power of natural capitalism	
6. Awareness of sustainable development, BoP & global risk as significant challenges		

robust. Potential sustainable competitive advantages can accrue to the business as a consequence of this augmented natural-resource-based view. Without this worldview, however, the natural environment could create a serious constraint on firms' attempts to create sustainable advantage.

These arguments are of equal import to both developing countries and the developed world because corporate sustainability in emerging markets is less embedded in corporate strategies, less pervasive and less politically rooted than in most high-income OECD countries. Notwithstanding the above, sustainable development is a critical challenge for both emerging and developed economies.

Organizational leadership plays a significant role in determining how this critical challenge is met. Using insights from leadership of social movements, contingency theories of leadership during uncertain times, emerging sustainability leadership lessons from praxis and the initial findings of an on-going neo-Glaserian study, a Zeitgeist Leadership Repertoire – Integrating Cognition, Conscience and Collective Spirit is offered as a set of leadership virtues, with their associated practices and related enactments. While research is continuing, the classification appears promising given its versatility in describing plausible leadership responses to prevailing organizational challenges.

REFERENCES

Aragon-Correa, J.A. and S. Sharma (2003), 'A contingent resource-based view of proactive corporate environmental strategy', *Academy of Management Review*, **28** (1), 71–88.

Arora, B. and R. Puranik (2004), 'A review of corporate social responsibility in India', *Development*, **47** (3), 93–100.

Balasubramanian, N.K., D. Kimber and F. Siemensma (2005), 'Emerging opportunities or traditions reinforced? An analysis of the attitudes towards CSR, and trends of thinking about CSR, in India', *The Journal of Corporate Citizenship*, **17** (Spring), 79–93.

Barney, J.B. (1991), 'Firm resources and sustained competitive advantage', *Journal of Management*, **17** (1), 99–120.

Barney, J.B. (2001), 'Resource-based theories of competitive advantage: a ten-year retrospective on the resource-based view', *Journal of Management*, **27** (6), 643–51.

Baskin, J. (2006), 'Corporate responsibility in emerging markets', *Journal of Corporate Citizenship*, **24** (Winter), 29–47.

Bent, D. and S. Draper (2007), *Profitable Today, Sustainable Tomorrow*, http://businessinnovationfacility.org/forum/topics/leader-business-strategies, accessed 20 January 2011.

Benyus, J.M. and D. Baumeister (2002), 'Packaging tips from the porcupine fish (and other wild packagers)', *Whole Earth*, **110** (Winter), 26–9.

Boyle, M. and J. Boguslaw (2007), 'Business, poverty and corporate citizenship: naming the issues and framing solutions', *Journal of Corporate Citizenship*, **26** (Winter), 101–20.
Chapple, W. and J. Moon (2005), 'Corporate social responsibility in Asia: a seven country study of CSR web site reporting', *Business and Society*, **44** (4), 415–41.
Christensen, C.R., K.R. Andrews, J.L. Bower, G. Hammermesh and M.E. Porter (1982), *Business Policy: Text and Cases*, 5th edn, Homewood, IL: Irwin.
Crews, D.E. (2010), 'Strategies for implementing sustainability: five leadership challenges', *S.A.M. Advanced Management Journal*, **75** (2), 15–22.
Crutzen, P.J. (2009), 'Can we survive the "anthropocene" period?', http://www.project-syndicate.org/commentary/crutzen1/English, accessed 16 January 2012.
Dierickx, I. and K. Cool (1989), 'Asset stock accumulation and sustainable competitive advantage', *Management Science*, **35** (12), 1504–11.
Duin, A.H. and L.L. Baer (2010), 'Shared leadership for a Green, Global, and Google world', *Planning for Higher Education*, 30–38.
Epstein, M.J, Buhovac, A.R. and K. Yuthas (2010), 'Implementing sustainability: the role of leadership and organisational culture', *Strategic Finance*, 41–7.
Faber, N., R. Jorna and J. van Engelen (2005), 'The sustainability of "sustainability" – a study into the conceptual foundations of the notion of "sustainability"', *Journal of Environmental Assessment Policy and Management*, **7** (1), 1–33.
Fletcher, J.K. and K. Kaufer (2003), 'Shared leadership: paradox and possibility', in C.L. Pearce and J.A. Congers (eds), *Shared Leadership: Reframing the Hows and Whys of Leadership*, Thousand Oaks, CA: Sage, pp. 21–47.
Gladwin, T. (1992), 'The meaning of greening: a plea for organisational theory', in K. Fischer and J. Schot (eds), *Environmental Strategies for Industry*, Washington, DC: Island Press, pp. 37–62.
Gladwin, T.N., J.J. Kennelly and T. Krause (1995), 'Shifting paradigms for sustainable development: implications for management theory and research', *The Academy of Management Review*, **20** (4), 874–907.
Glaser, B.G. and A.L. Strauss (1967), *The Discovery of Grounded Theory: Strategies for Qualitative Research*, Chicago: Aldine.
Hall, C. (2007), 'Are emerging-market TNCs sensitive to corporate responsibility issues?', *Journal of Corporate Citizenship*, **26** (Summer), 30–36.
Hart, S.L. (1995), 'A natural-resource-based view of the firm', *Academy of Management Review*, **20** (4), 986–1014.
Hart, S.L. and G. Dowell (2011), 'A natural resource-based view of the firm: fifteen years later', *Journal of Management*, **37** (5), 1464–79.
Heifetz, R.A. and D.J. Laurie (1997), 'The work of leadership', *Harvard Business Review*, **75** (1), 124–34.
Hind, P., Wilson, A. and G. Lenssen (2009), 'Developing leaders for sustainable business', *Corporate Governance*, **9** (1), 7–20.
Hoffman, A.J. (2007), 'If you are not at the table, you are on the menu', *Harvard Business Review*, **84** (10), 34–5.
Howard-Grenville, J.A. (2006), 'Inside the "black box": how organisation culture and subcultures inform interpretations and actions on environmental issues', *Organisation & Environment*, **19** (1), 46–73.
Karnani, A. (2007), 'The mirage of marketing to the bottom of the pyramid: how the private sector can help alleviate poverty', *California Management Review*, **49** (4), 90–111.

Kegan, R. and L. Lahey (2010), 'Adult development and organisational leadership', in N. Nohria and R. Khurana (eds), *Handbook of Leadership Theory and Practice: A Harvard Business School Centennial Colloquium*, Boston, MA: Harvard Business Press, pp. 769–87.

Khan, M.N. and M.N. Faisal (2001), 'Mobile commerce: the next dimension', *IIMB Management Review*, **13** (4), 7–18.

Kleiner, A. (1991), 'What does it mean to be green?', *Harvard Business Review*, **69** (5), 38–47.

Lovins, A.B., L.H. Lovins and P. Hawken (2007), 'A road map for natural capitalism', *Harvard Business Review*, **85** (7/8), 172–83.

Lubin, D.A. and D.C. Esty (2010), 'The sustainability imperative', *Harvard Business Review*, **88** (5), 42–50.

McDonough, W. (2005), 'Ecology, ethics and the making of things', *Sojourners Magazine*, **34** (5), 36–8.

McDonough, W. (2009), 'Key note address: business as an agent of world benefit', http://www.youtube.com/watch?v=DogArXwGWlI, accessed 10 January 2010.

McDonough, W., M. Braungart, P.T. Anastas and J.B. Zimmerman (2003), 'Applying the principles of green engineering to cradle-to-cradle design', *Environmental Science and Technology*, **37** (23), 434–41.

Murthy, V. and D. McKie (2009), *Please Don't Stop the Music: An Ensemble Leadership Repertoire, Productive Sustainability, and Strategic Innovation for Uncertain Times*, London, UK: Inderscience.

Nelson, J. (2003), *Economic Multipliers: Revisiting the Core Responsibility and Contribution of Business to Development*, London: International Business Leaders Forum (IBLF).

Nidumolu, R., C.K. Prahalad and M.R. Rangaswami (2009), 'Why sustainability is now the key driver of innovation', *Harvard Business Review*, **87** (9), 56–64.

NSE/NIFTY (2003), 'Proposal 84: corporate social responsibility of NSE nifty companies, content, implementation strategies, and impact', NSE Research Papers, Mumbai: National Stock Exchange.

Pearce, C.L., C.C. Manz and H.P. Sims Jnr (2009), 'Where do we go from here? Is shared leadership the key to team success?', *Organisational Dynamics*, **38** (3), 234–38.

Porritt, J. (2006), *Capitalism as if the World Matters*, London: Earthscan.

Porter, M.E. and M.R. Kramer (2006), 'Strategy and society: the link between competitive advantage and corporate social responsibility', *Harvard Business Review*, **84** (12), 78–92.

Porter, M.E. and F.L. Reinhardt (2007), 'A strategic approach to climate', *Harvard Business Review*, **85** (10), 22–4.

Prahalad, C.K. and G. Hamel (1990), 'The core competence of the corporation', *Harvard Business Review*, **68** (3), 79–91.

Prahalad, C.K. and S.L. Hart (2002), 'The fortune at the bottom of the pyramid', *Strategy+Business*, **26** (1st quarter), 54–67.

Quinn, L. and M. Dalton (2009), 'Leading for sustainability: implementing the tasks of leadership', *Corporate Governance*, **9** (1), 21–38.

Reserve Bank of India (2009), 'Monthly bulletin, Table 44: foreign exchange reserves', http://rbidocs.rbi.org.in/Bulletin/PDFs/89876.pdf, accessed 3 January 2011.

Rudd, K. (2008), 'Leadership for long term sustainability: the roles of government, business and the international community', address to the National Business

Leaders Forum on Sustainable Development, Parliament House, Canberra, 19 May, http://www.pm.gov.au/media/Speech/2008/speech_0253.cfm, accessed 22 September 2008.
Sachs, J.D. (2008), 'A user's guide to this century', *The National Interest*, **96** (July/August), 8–14.
Sachs, J.D. (2010), 'Globalisation in the era of environmental crisis', *International Trade Forum*, **1**, 7–10.
Sankaran, K. (2003), 'Imperatives of a civil corporation in the new economy', *IIMB Management Review*, December Issue, 82–8.
Selznick, P. (1957), *Leadership in Administration: A Sociological Interpretation*, Evanston, IL: Peterson.
Sharma, S. (2003), 'Research in corporate sustainability: what really matters?', in S. Sharma and M. Starik (eds), *Research in Corporate Sustainability: The Evolving Theory and Practice of Organisations in the Natural Environment*, Cheltenham, UK and Northampton, MA, USA: Edward Elgar, pp. 1–29.
Siegel, D.S. (2009), 'Green management matters only if it yields more green: an economic/strategic perspective', *Academy of Management Perspectives*, **23** (3), 5–16.
Teece, D.J., G. Pisano and A. Shuen (1997), 'Dynamic capabilities and strategic management', *Strategic Management Journal*, **18** (7), 509–34.
van Marrewijk, M. (2003), 'Concepts and definitions of CSR and corporate sustainability: between agency and communion', *Journal of Business Ethics*, **44** (2–3), 95–105.
Visser, W. (2008), 'Corporate social responsibility in developing countries', in A. Crane, A. McWilliams, D. Matten, J. Moon and D. Siegel (eds), *The Oxford Handbook of Corporate Social Responsibility*, Oxford: Oxford University Press, pp. 473–9.
Wernerfelt, B. (1984), 'A resource-based view of the firm', *Strategic Management Journal*, **5** (2), 171–80.
Wilkinson, L. (2009), 'How to build scenarios', http://www.wired.com/wired/scenarios/build.html, accessed 16 January 2012.
World Bank (2006), 'Development and the next generation', *World Development Report 2007*, Washington: World Bank.
WEF (World Economic Forum) (2012), 'Global risks 2012 (7th edn): an initiative of the risk response network', http://www3.weforum.org/docs/WEF_GlobalRisks_Report_2012.pdf, accessed 16 January 2012.

12. The Asian century is the Chinese century and the century of turbulence

Malcolm McIntosh

INTRODUCTION

What is described as the Asian century may more accurately be described as the Chinese century or the BRIICS century (Brazil, Russia, India, Indonesia, China and South Africa) as it these countries that are in the economic ascendant. But however the century is described geographically it will also be the age of turbulence and disruption. This is because we have entered the Anthropocene age: the age where we must learn to live as one race, on one planet within planetary limits, or we may not see the next century.

According to the current debate there is a transfer of economic and political power back to Asia after some 500 years of dominance by the European model. Most latterly this has been driven by the United States of America's neo-liberal market-dominated model. Across Asia the lack of homogeneity and the vast disparities that exist in these regions begs the question of what is meant by 'Asia'. To compare Asia to Europe or North America may be to mix apples and pears, or persimmons and lychees.

The idea of an Asian century requires further discussion and deconstruction because it may be an illusion. It is not an homogenous region at all, and the Asian attribution may be a post imperial feeling of 'it's our turn again' or 'it's not the American century, that's so last century' or 'it's not the European century, that's so nineteenth century'. Let us not forget that in the 1960s it was 'feared', particularly in the USA, that we were entering the Japanese century as their economy's upward trajectory seemed unstoppable. Of course it wasn't 'feared' in Japan, it was celebrated. Similarly, sometime in the second half of the last century there was talk of the Pacific century, and, looking at the current rise of the BRIICS countries, this had some validity.

The other problem of talking about the Asian century is that the Asia that is referred to outside Asia is a much misunderstood monolith in Europe and North America. As many commentators have pointed out there is a real fear of the unknown, as most non-Asians have little or no understanding of Asia, per se, and fail to distinguish between China and India, and between Japan, Taiwan and Singapore despite the fact that they are historically, culturally, economically and politically very different. This is not dissimilar to non-Europeans thinking that Sweden is the same as Greece, or Spain similar to Croatia. The difference between seeing Europe and Asia as homogenous regions is that the former have formed a political and economic union or community, whereas there are many economic and political economies across Asia.

We live in a globalized world and the deterritorialisation that has come about through boundaryless social media and loosely regulated markets means that the way we see the world and the fundamental characteristics of globalization must be re-examined, particularly in relation to Asian and non-Asian perspectives.

THE ANTHROPOCENE

'The Asian century' needs to be understood in the context of the Anthropocene – an understanding that the Earth's lithosphere has changed significantly because of humanity's impact.[1] The Anthropocene is generally agreed to have started with the era of mass industrialisation, in other words over the last two hundred years. It is hard to imagine a world again that does not need to be managed, such is the mismanagement by humans over the last few hundred years. On one hand it is time to celebrate our success as a species, as a single race. Our ingenuity, creativity and problem solving abilities have given the majority of the population – which continues to grow exponentially – long, happy, contented lives allowing us to travel at speed and visit faraway places on metal birds; to lie in the sun drinking chilled pinot grigio; to call the rescue services when our personal systems go down.

We know so much about the world; in fact we are awash with information – but what do we do with all that we know? Just as we think we have a clear picture of the world with our personal mental models of PlanetHome,[2] we are jolted into a new reality by incoming data which confounds our placidity and pacifity. Humans have always had to learn to survive but it is hard in the Anthropocene to cling to the life raft of personal identity (who am I, where do I come from, where do I live?), and come to terms with our responsibilities of global citizenship (what can I do about the world and what is my place in it?), and to meditate on the existential angst (what does it mean to be human now that we know what we know about the state of spaceship Earth?).[3]

These questions are no less relevant to Asian residents of Tokyo, Jakarta, Mumbai, Suva and Lhasa than they are to non-Asians in New York, London, Paris or Cape Town. Multi-stakeholder roundtables convened by this author in 2008/9 in the cities of London, New York, Sydney, Toronto, Beijing and Cape Town showed the same four issues arose in relation to the necessary transition to a sustainable enterprise economy in each of these locations: the lack of scientific knowledge amongst the general population that would enable most people to understand climate change science and global warming; institutional inertia preventing the necessary radical change in society; a political leadership deficit which means decisions are too often made with a short term opportunity in mind rather than a long term perspective; and, a lack of democratic participation in issues that require either a long term or a global governance perspective. A combination of all four factors was seen in five continents as holding back the necessary transition to a low pollution, Earth-centric economy (McIntosh and Waddock 2010). That these were the overall themes that emerged is salutatory and reemphasises the point that it is the whole of humanity that is faced with the challenge of change. This may be the Asian century, but unlike previous major tectonic shifts in geopolitical influence and power, this time what is American is Asian is what is African is European, and vice versa. We are all in this together, on Spaceship Earth, Asian century or Global century.

Calls for a new global political economy, for global citizenship and responsible leadership were highlighted in the High Level Panel report on *Global Sustainability: Resilient People, Resilient Planet* presented to the UN (United Nations) Secretary-General prior to the Rio+20 meetings in June 2012. The report called for a 'new political economy' which brings together economists, social activists and environmental scientists. This means that there should be no difference between ecology and economics, they do after all have the same Greek etymology: 'ecos', meaning home, now PlanetHome. The report also highlighted the fact that the world is experiencing the best and the worst of times: unprecedented prosperity on one hand for some people and unprecedented stress for the planet on the other hand. Arguing for hope and the triumph of human ingenuity over disaster the report says that 'we are not passive, helpless victims of the impersonal, deterministic forces of history. The exciting thing is that we can choose our future'.[4] Other commentators have adopted a similar stance faced with the apocalyptic possibilities presented by climate change. Tim Flannery published a book subtitled 'an argument for hope' and others, such as Paul Gilding, are talking up the possibility of disruption followed by a new economy. They are in contrast to writers such as Slovak Zizek (2011) and Clive Hamilton (2010) who write of 'the end times' and 'a requiem for a species'. One thing is certain though, and that is that it may be in Asian economies that the ideas and technologies, and perhaps the economic and political models, will come to enable humanity to face the next great transition to the sustainable enterprise economy.

DEFINING THE REGION

It is not easy deciding what is meant geographically by the Asian century and it may be that it is a term invented by a frightened non-Asian world that wants to put all Asian countries in the same basket. As the Director of Griffith University's Asia Institute, Andrew O'Neill, says: 'you decide what is in or out, as there is no accepted definition'.[5]

One comment on defining the region is that 'it depends where you start from'. Many Asians like the idea of being seen by the rest of the world as the centre of attention, affluence and influence, some see their particular country as the centre of Asia, and others know that this blanket description is a problem as it defines all Asians as the same and applies an homogeneity to a region that does not exist with the blind specificity that is perceived outside the region.[6]

Let us be content with the amorphous understanding of Asia in the context of the Asian century and accept that it is fundamentally concerned with a very significant shift to some Asian countries economically, and that this started at the end of the second world war with the rise of the Asian tigers, as they were called, led by Japan. In recent years it has been followed by the economies of China, India and Indonesia growing in terms of trade and improvements in average per capita wealth such that the region, however it might be defined, now dominates the global economic landscape. Two things are to be remembered. First that average wealth in countries with such extreme inequality as India and China is an inadequate measure of progress towards what are regarded as civilized societies in terms of indexes such as the UNDP's HDI.[7] Second, it must not be forgotten in the rush to ascribe this century to Asia that the German and US economies are also mighty and that the US remains larger than all and that it alone dominates global security in terms of militarisation and forward projection. In 2012 the US is the world's policeman, able to rattle a significant stick at all who threaten its hegemony. But, this is changing rapidly and even though, or because, the US has pronounced its military strategy to have equal focus on the Atlantic and the Pacific, the rise of China as a military power continues apace. This is not because China is sabre rattling or significantly aggressive in its posture but simply because it wishes to be able to protect its supply chains from Europe, the US and the Middle East and because it is geographically and economically a large power.

ASIAN CENTURY CONTRASTS: AUSTRALIA, CHINA AND INDIA

In the 2012 White Paper 'Australia in the Asian Century' the Australian government divides the world between West and East: 'Western economies are defined as: Austria, Belgium, Canada, Denmark, Finland, France, Germany, Italy, the Netherlands, Norway, Sweden, Switzerland, the UK and the US. Asian economies are defined as: Bangladesh, China, Hong Kong, India, Indonesia, Japan, Republic of Korea, Malaysia, Pakistan, the Philippines, Singapore, Sri Lanka, Taiwan, Thailand and Vietnam'.[8] Quite correctly, but paradoxically, Australia and New Zealand are excluded from both lists. It may not be widely recognised in either country but their histories increasingly make them distinct from Europe, North America and Asia: they are what they are – not European, North American or Asian.

Let us briefly look at three contrasting places apparently in the Asian region. First we take two contrasting cities in two different hemispheres in Australasia that will double their populations, infrastructure, economies and wealth in the next few decades at current projections and that are bound together by trade.

One is Brisbane in subtropical Queensland in Australia, whose wealth is largely derived from its resource trade with Asia, most particularly with China, South Korea and Japan. The other is Shanghai which is doing its best to emulate its near neighbour Tokyo (even if it won't openly admit it) – which boomed in the second half of the last century.

In 2012 both cities, Brisbane and Shanghai, are experiencing dramatic economic growth and are addicted to concrete and road building. Brisbane is capital of Queensland, the largest state in Australia, where the state government is the third largest land owner in the world after Russia and China: the state is the size of France and Germany together. Australia is the size of Europe or the USA and yet it has just 22 million people, 85 per cent of whom live within an hour of the coast. It's the driest and oldest continent on the planet and packed with mineral reserves. It is a minnow to China's 1.4 billion people and has no cities near the size of Shanghai.

Australia has recently been referred to as 'China's mine' because of the latter's exploitative relationship with a country that has most of the mineral resources that China needs for its massive industrialisation and economic development. Both are in the Asia Pacific region; both are Australasian countries; both are contributing to the Anthropocene. In one sense they epitomise the Asian century, even though they are poles apart: the distances, the different geographies, the triumphs, the disasters, the disparities, the inefficiencies, the political systems and the trade. Other contrasts are also compellingly interesting.

China has had a one child policy for many decades, which means that there are about one billion fewer people on the planet. Australia by comparison has a policy whereby parents are rewarded with an AU$5000 baby bonus because they are trying to significantly increase the local population rather than encourage immigration.

Further contrasts highlight the differences in the region. Australia is one of the most egalitarian and democratic countries in the world, ranking second in the UNDP's (United Nations Development Programme) Human Development Index[9] and sixth in the Economist's Democracy ranking,[10] and, apart from its treatment of Indigenous people, has one of the best Gini coefficients in the world, while China has dramatically rising wealth inequality not dissimilar to the USA.[11]

In another example of the disparities in the Asia region, Amartya Sen, Nobel Prize winning former World Bank economist, argues that a strange paradox has arisen in terms of models of successful development. When it comes to progress on human security and sustainable development issues, Sen argues that China's one party totalitarianism beats Indian democracy. For example:

> The Chinese authorities . . . . from 2004 rapidly started reintroducing the right to medical care. China now has a considerably higher proportion of people with guaranteed health care than does India. The gap in life expectancy in China's favor has been rising again, and it is now around nine years; and the degree of coverage is clearly central to the difference. Whether India's democratic political system can effectively remedy neglected public services such as health care is one of the most urgent questions facing the country.[12]

The Asian century poses paradoxes and challenges to most models of development, not least in challenging market liberalisation but also in putting at the top of the agenda the link between democratic participation, social justice, living within the earth's carrying capacity, and market behaviour. If the last five hundred years are challenged it may not be in terms of market systems, for these are being adopted across Asia, but in the models of participatory, democratic government. Of particular concern are levels of wealth disparity, corruption and the absence of equality of treatment before the law. The European social welfare model may be threatened not just by over indebted European governments but by rampant neoliberal models which are happy to see significant sections of populations live in poverty and struggle for survival on the grounds that for markets to work well there must be cheap labour and a sense of Darwinian struggle. Collectivism as represented by the European or Australia and New Zealand model, where 'leave no one behind' is a general but not absolute tenet, could be threatened by less inclusive models founded on US principles of the right to property, life, liberty and happiness.[13]

CORPORATE RESPONSIBILITY AND THE ASIAN CENTURY

For countries in Asia the ten year development of the international guidance on corporate social responsibility (ISO 26000) has been an opportunity to engage with the many market-based corporate social responsibility (CSR) initiatives that have arisen since 1990. The countries in the region that sent representatives to participate included China, India, Indonesia, Mongolia, Australia and others. A number of prominent non-state organizations were also represented and their multi-stakeholder CSR initiatives are listed in the standard's annex. As most of these initiatives had originated in, or been driven or financed by, non Asian countries it was salutory to see such participation in global standard setting on corporate responsibility by countries such as China, Japan and India.

BRIIC companies now occupy 50 of the top 100 world's largest companies and of that top 100, 19 are state owned enterprises, mostly from Brazil, Russia and China (*The Economist* 2011; Cohen 2011). The inclusion of these Asian countries and companies and international civil society organisations was crucial to the evolving development of new global governance mechanisms that go beyond simple multinational negotiation. It is difficult for some in the West to understand the shift that is taking place not just in the evolving new global governance space but also in the role of neo-capitalist organisations that operate in the market but are arms of the state, many of them based in Asia.

While a US President may call on US companies to create more jobs for Americans, the US Department of Commerce says that US companies did create plenty of jobs in the last decade: 683 000 in China and 392 000 in India for instance. A similar story can be told of the Australian logistics company Linfox who in the year March to March 2010–11 created 500 jobs in Australia and 3500 in South East Asia.[14] The point is that Asian companies, by virtue of a mixture of domicility, loyalty, identity and state ownership, are more likely to create jobs in what they see as their own backyards than US and European companies who have fewer loyalties or ties to government and are essentially supra-territorial corporations loyal to no one state (even if they do expect to be bailed out by a nominal host state when they go bust as in the case of the Royal Bank of Scotland in the UK in 2008).

ISO 26000, a guidance standard on corporate responsibility released in November 2010, sets standards for companies across the whole of the Asian and Asia Pacific region as well as the rest of the world, and the fact that it was negotiated with due deliberation and participation from members' governments is a positive sign for new global governance that begins to address the most pressing international issues, and in a way that is more inclusive than previous mechanisms.[15] Interestingly, in the context of this chapter on sustainability and responsibility, the standard also refers to sustainable business:

> Sustainable business for organizations means not only providing products and services that satisfy the customer, and doing so without jeopardizing the environment, but also operating in a socially responsible manner.[16]

Further to that it provides a new useful refinement of the definition of corporate responsibility, which, given that many companies in the Asia region are new and emerging, that many are state owned and controlled, and that many are emerging from non-market situations, is enormously powerful:

> The Social Responsibility of an organisation is for the impacts of its decisions and activities on society and the environment, through transparent and ethical behaviour (that) contributes to sustainable development, including health and the welfare of society . . . takes into account the expectations of stakeholders . . . and is in compliance with applicable laws and norms of behaviour . . . (and) is integrated throughout the organisation in its relationships.[17]

Furthermore ISO 26000 is useful for providing another summation of the definition of sustainable development and sustainability which is important in the context of the rise of vast inequality and unsustainable practices in the Asia and Asia Pacific regions. Halina Ward examines the ISO 26000 definition of sustainable development and notes that.

> Sustainable development is about integrating the goals of a high quality of life, health and prosperity with social justice and maintaining the earth's capacity to support life in all its diversity. These social, economic and environmental goals are interdependent and mutually reinforcing. Sustainable development can be treated as a way of expressing the broader needs of society as a whole. (Ward 2012, p. 38)

The development of ISO 26000 within a global governance framework is a significant advance from the multiplicity of voluntary multi-stakeholder CSR initiatives that have taken root over the last decade of the twentieth century and the first decade of the twenty-first century. It begins to fill a gap that the UN and the WTO have failed to address adequately, namely how will the emerging sustainable enterprise economy look? How will the UN High Level Group's desire for a new political economy be shaped and what will be its characteristics, standards and norms?

In this way ISO 26000 adds a fourth and perhaps a fifth leg to the three legs of sustainability. Along with social, economic and environmental legs we now have a more stable stool with two more legs: governance, and, progressively culture. Institutional inertia, the democratic deficit and the leadership challenge are being met in the first steps to global agreement on new market capitalism in the Asian century.

Amongst the 450 participants and 210 observers from 99 International Organization for Standardization (ISO) member countries were representatives of 42 other organisations including those UN groups and NGOs (non-government organisations) who had developed significant multi-stakeholder standard setting, certification, monitoring and engagement programmes over the previous two decades. While ISO 26000 might be described as the lowest common denominator it is already taking its place as an authoritative source of definitions and guidelines on a global basis. Its credibility partially rests on the fact that it is not Western in origin, and in particular that it is a post-colonial document with developed and developing world perspectives. This is most important and useful in the context of the Asian century.

Research carried out at the Asia Pacific Centre for Sustainable Enterprise at Griffith University has sought to examine the links between national sustainable development indicators and the adoption of CSR initiatives in those countries. One of the objects of this ongoing research exercise is to see if there is a link between market and national political drivers, but this research will also in future years indicate how different ISO standards and guidelines drive processes. For instance, does the adoption of ISO 9000 in India drive the referencing of ISO 26000, or the adoption of ISO 14000 in Japan increase the chance of ISO 26000 becoming a common benchmark in Japan? Similarly from our research work, is there any correlation between the number of signatories to the UN Global Compact and the adoption of ISO 26000 standards?

Table 12.1 is a summary chart of organisational indicators for the Asia Pacific region which was drawn up in 2009–11. Countries are ranked first by GDP (as it was in 2009), then by population, and then by the corporate take up of a number of CSR initiatives: UN Global Compact signatories,

Table 12.1 Summary chart for organisational indicators: Asia–Pacific region (by GDP 2009)

Country	GDP 2009 US$ mill.	Population mid-2009 (mill.)	UNGC[a]	GRI[b]	ISO9001 (December 2008)	ISO14001 (December 2008)	SA8000
1 Japan	5067526	127.6	145	89	62746	3573	2
2 China	4909280	1331.4	220	52	224616	39195	349
3 Australia	924843	21.9	88	72	8773	1125	0
4 Korea, Republic of	832512	48.7	190	54	23036	7133	3
5 Indonesia	540277	243.3	132	1	5713	849	9
6 Thailand	263856	67.8	28	3	5275	934	14
7 Hong Kong	215355	7.0	15	0	3499	739	20
8 Malaysia	191601	28.3	71	6	6267	997	2
9 Singapore	182232	5.1	73	8	4528	668	1
10 Philippines	160476	92.2	47	9	2412	773	5
11 New Zealand	125160	4.3	4	10	1228	148	0
12 Vietnam	91854	87.3	94	0	3971	325	48
13 Cambodia	10447	14.8	0	0	7	6	0

14 Papua New Guinea	7893	6.6	0	1	24	4	0
15 Laos	5939	6.3	1	0	2	0	0
16 Solomon Islands	657	0.5	0	0	0	0	0
17 Timor Leste	558	1.1	1	0	0	0	0
18 Tonga	311	0.1	0	0	0	0	0
19 Burma	n.a.	50.0	0	0	18	843	0
Total	13 530 777	2 144.3	1 109	306	352 113	89 312	453
Average	751 710	112.9	58	16	18 532	4 701	24

Notes:

1st
2nd
3rd–8th
9th–15th
16th–19th

a. UN Global Compact: data as of May 2011.
b. GRI (Global Reporting Initiative) for 2009, Data as of August 2010.
c. Data as of 31 December 2010.

Source: Forbes et al. (2012).

the adoption of GRI (Global Reporting Initiative), ISO 9001, ISO 14001, and SA 8000. The reason why ISO 26000 is so significant to the ongoing research work is because it does what none of the CSR initiatives do: it marries *corporate* and *national* thinking, standards, norms and guidance on sustainability and corporate responsibility. In the context of the Asia and Asia Pacific regions this is significant because within those regions there are all manner of governmental systems. These range from the world's largest democracy, India, to one of the world's most democratic countries, Australia, to the world's largest country under a single party communist non-democratic regime, China, to the world's largest Muslim democracy, Indonesia, to a country, at the time of writing, run by a military junta, Myanmar/Burma.

As we have seen, many of the signatories to the UN Global Compact in China are some of the largest companies in the world and constitute many of the nineteen companies out of the top one hundred corporate economies that are also state owned enterprises. These companies operate on a global basis and have found that their adoption of the ten principles of the UN Global Compact has put them in quasi global governance situations. This is no bad thing as long as it leads to the development of more politically sensitive companies and board room directors who understand that corporate responsibility has moved from 'know what you do', to 'do no harm', to 'take responsibility'.

A good example of the paradoxes and dilemmas for companies is found in the Chinese company Sinopec, which upholds the single child policy when operating in China but also has to observe the current polygamy consensus when operating in South Africa. Given the Global Compact principles, the ISO 26000 guidelines and embedded and explicit governance components of ISO 26000, to which the Chinese and South African governments contributed, how should this Chinese company operate in different geographical and cultural contexts?

Let us suggest that this example highlights the paradox at the very heart of the idea of the Asia century. First, it is not hegemony in the way that European colonialism was, or the US century has been. Second, Asia contains within it so many differences as to make the idea of homogeneity unsound. Third, Asia is so diverse that to think of the current Chinese developmental progress as comparable to Japanese post-war development or the same as Singapore's development may be stretching an idea to the point of absurdity. Fourth, the differences in socio-economic development across the Asia and Asia Pacific regions are extraordinary but they mirror similar socio-economic differences in other parts of the world, not least in South America and Africa.

The increasing wealth disparities in the USA and in some European countries also illustrate the point that the world is severally layered: by socio-economic groupings; by access to and use of social media; by functioning and non-functioning rule of law jurisdictions; and by the behaviour, role, scope and purpose of supraterritorial corporations such as Chinese state owned Sinopec, Anglo–Dutch registered Unilever, and Indian owned and driven Tata Industries, in different political economy settings.

In short the Asian century may be an illusion. The reality is that while the Asian Tigers in the period 1945–2000 shook the world, it is because of the size of the Chinese and Indian economies that the world outside Asia is calling this the Asian century. Further, the Asian century is also the BRIICS century, and most importantly it is the century of sustainability: we must learn to live within the Earth's carrying capacity or perish, we must curb population growth, we must devote our science, creativity and innovation to staying alive. To this end the Global Sustainable Competitiveness Index, which balances social cohesion, sustainable innovation and competitiveness, resource intensity and efficiency, and natural capital on an aggregate ranking out of 186 shows a wide range within the Asia and Asia Pacific regions: Japan is 9th; Australia 23rd; China 36th; Indonesia 61st; India 100th. A BRIICS perspectives puts Brazil at 25 and Russia at 56.[18]

The idea of the Asian century, as we said, depends where you start and how you want to see the world. But from any perspective it is myopic and self-referential and depends on what you want to see – and perhaps where you want to belong.

NOTES

1. The term 'anthropocene' is variously used by natural and social scientists alike. For instance see Flannery (2010) and Zizek (2011).
2. I first used this term in *PlanetHome* in Shah et al. (2003).
3. *Spaceship Earth* is a term made popular by Barbara Ward in 1966 in a book of the same name. It was much used in the 1960s.
4. See UN High Level Panel Report on *Global Sustainability: Resilient People, Resilient Planet*, 30 January 2012, New York.
5. Personal communication, 2011.
6. See Bendell and Ng (2009).
7. For instance see 'Europe is still the closest thing to paradise on Earth despite the Eurozone crisis, Europeans should not forget how great a place to live their continent is', Remi Adekoya, guardian.co.uk, Wednesday 19 September 2012.
8. See http://asiancentury.dpmc.gov.au/issues-paper/what-is-happening-in-asia.
9. See http://hdr.undp.org.
10. See http://eiu.com/DemocracyIndex2011.
11. Gini coefficient table, http://data.worldbank.org/indicator/SI.POV.GINI.
12. See http://www.nybooks.com/articles/archives/2011/may/12/quality-life-india-vs-china.

13. For more on the past and the coming debate see Judt (2010) and Crouch (2011).
14. Lindsay Fox quoted in the *Australian Financial Review*, 31 March 2012, p. 3.
15. For discussion, see Ward (2012).
16. http://www.iso.org/iso/iso_catalogue/management_and_leadership_standards/social_responsibility/sr_iso26000_ overview.htm, 21 March 2012.
17. Ibid.
18. http://www.weforum.org/issues/global-competitiveness.

REFERENCES

Bendell, Jem and Chew Ng (2009), *Characteristics of Asian CSR*, Lien Centre.
Cohen, Roger (2011), 'The corporate divide', *New York Times* and *The Observer*, 11 December.
Crouch, Colin (2011), *The Strange Non-Death of Neoliberalism*, Cambridge: Polity.
Flannery, Tim (2010), *Here on Earth*, Melbourne: Textpublishing.
Forbes, Susan, Malcolm McIntosh and Andrea Haefner (2012), 'Towards CSR and the sustainable enterprise economy (2) in the Asia Pacific region', September, Asia Pacific Work in Progress Papers Series, http://www.griffith.edu.au/business-commerce/asia-pacific-centre-for-sustainable-enterprise/publications/working-paper-series/issue-1.
Hamilton, Clive (2010), *Requiem for a Species*, Australia: Allen and Unwin.
Judt, Tony (2010), *Ill Fares the Land*, London: Penguin.
McIntosh, Malcolm and Sandra Waddock (2010), 'Learning from the roundtables on sustainable enterprise', in Andreas Rasche and Georg Kell (eds), *The United Nations Global Compact: Achievements, Trends and Challenges*, Cambridge: Cambridge University Press.
Shah R., Murphy, D. and M. McIntosh (2003), *Something to Believe In*, Sheffield: Greenleaf Publishing.
The Economist (2011), 'The company that ruled the world', 17 December.
Ward, Halina (2012), 'ISO 26000 and global governance for sustainable development', London: IIED.
Zizek, Slavoj (2011), *Living in the End Times*, London: Verso.

Index

Titles of publications are in *italics*.